Textbook of PTERIDOPHYTES

THE AUTHOR

Dr. Inderdeep Kaur is a Professor in Botany, Department of Botany, Sri Guru Tegh Bahadur Khalsa College, University of Delhi, Delhi. She has an experience of teaching Archegoniates to undergrads for more than fifteen years. Besides teaching, she retains interest in research and has guided PhD students and handled independent projects funded by UGC and DST. The outcome of research endeavours has fetched her publications in international and national journals of repute. She has presented her work in several conferences at international levels. She has contributed to the syllabus revision at University of Delhi during FYUP, CBCS and LOCF based curriculum at undergraduate level. Her recent books – **Textbook of Bryophytes** and **Textbook of Gymnosperms** have been received well both by faculty and students. She also has to her credit a co-authored book -**Brown Algae: Structure, Ultrastructure and Reproduction** and a co-edited book -**Plant Reproductive Biology and Conservation**.

Textbook of PTERIDOPHYTES

Based on the syllabus of New Education Policy

INDERDEEP KAUR
Professor of Botany
Sri Guru Tegh Bahadur Khalsa College
University of Delhi, Delhi

2024

Daya Publishing House®

A Division of

Astral International Pvt. Ltd.

New Delhi – 110 002

ISBN 9789359191881

Published by : **Daya Publishing House®**
A Division of
Astral International Pvt. Ltd.
– ISO 9001:2015 Certified Company –
4736/23, Ansari Road, Darya Ganj
New Delhi-110 002
Ph. 011-43549197, 23278134
E-mail: info@astralint.com
Website: www.astralint.com

PREFACE

The past two decades have witnessed unprecedented interest and a revolution in our understanding of phylogenetic relationships of vascular plants. The role of Pteridophytes in evolution of present day vascular land flora is pivotal as their existence dates back to Silurian time with successful survival thereafter. Such an insight into the phylogenetic relationships has helped in construction of the tree of life of vascular plants.The taxonomy and classification of this group of archegoniates has always been in a state of flux. However, taxonomic treatments of Smith et al. (2006) and PPG1proposed by Schuettpelz et al. (2016) is an updated new information. These classifications are based on molecular phylogenetic analysis of ferns and lycophytes. The studies indicate that lycopods are monophyletic and are the earliest diverging group within vascular plants. *Psilotum* on the other hand, is now considered more closely related to the seed plants.

The style of presentation of this book is same as for author's earlier books - Textbook of Bryophytes and Textbook of Gymnosperms. The lucid account of taxa besides simple, and easy-to-learn presentation is the strong point of this book. There are thirteen chapters in all, with Chapter 1, An **Introduction** to the world of Pteridophytes. Chapter 2 deals with **Classification** where three recent trends are provided. These classifications which do not accept the term 'fern allies' but consider Lycopodiopsida and Polypodiopsida as two major classes, are discussed along with brief historical overview of classification. The author has adopted Schuettpelz et al. 2016 (PPG1) classification in the book. Chapter 3 starts with general features of ***Lycopodiopsida*** followed by the type study of ***Lycopodium***. The life cycle of ***Selaginella*** is detailed in Chapter 4 while Chapter 5 is a detailed account of the class - **Polypodiopsida** – the ferns. Chapters 6, 7, 8 and 9 deal with type studies of ***Psilotum*** (whisk fern), ***Equisetum*** (horsetails), ***Marsilea*** (four leaf clover) and ***Pteris*** (bracken fern) respectively. **Heterospory** which has played an important role in seed habit is detailed in Chapter 10 while **Stelar evolution**, a basis of vascular system of land plants is discussed in Chapter 11. The Chapters 12 and 13 deal with **Telome theory** and **Economic importance** respectively.

The book is provided with clear and well labelled diagrams at appropriate places in the text, making comprehension of the topic easy. The colour plates added at the end of the book would be of great help in perception of the architecture and organization of plants. Such photographs would help generate interest in practical classes related to the topics. A detailed glossary and questions based on the chapters would help students build up the concepts for entrance examinations. In addition, links to videos have been provided which would assist online teaching-learning.

I am grateful to Prof. Jaswinder Singh, Principal, Sri Guru Tegh Bahadur Khalsa College, University of Delhi, Delhi, for extending the college environment from where this scholarly insight emanated.

I am also immensely grateful to Prof. P.L. Uniyal for constructive comments on the recent developments in Pteridophytes.

Ms Preeti Giri, PhD scholar, Department of Botany, University of Delhi, extended her help in photography of the slides and this gesture is greatly appreciated.

I can't thank enough my family for the relentless encouragement and constant support which made it possible for me to compile this work during COVID -19 when everything almost came to a standstill.

Finally, I extend my gratitude to Astral International (P) Ltd. for the physical creation of this book.

Inderdeep Kaur

CONTENTS

1

CHAPTER

INTRODUCTION

Pteridophytes, the lower vascular plants are known to have originated in Silurian periods of Late Palaeozoic era, regarded as age of pteridophytes. The term pteridophyta has been derived from a Greek word Pteron meaning 'feather' and the name was originally used for the group because of the pinnate or feather-like fronds or leaves of the ferns. They are also called snakes of plant kingdom or botanical snakes because just as reptiles are the first land animals, pteridophytes are the first land plants. Before them, bryophytes were considered the only amphibian plants. The Psilophytopsida (earlier classification) are extinct pteridophytes while leptosporangiate ferns are the advanced forms. A few aquatic forms such as *Azolla, Salvinia* and *Marsilea* also exist. The primitive plants such as *Rhynia* and *Horneophyton* are extinct forms with evolutionary significance.

Lower vascular plants or the most primitive vascular plants, commonly known as pteridophytes, are also called vascular cryptogams. They are the spore-bearing vascular plants, and include the ferns, club mosses, spike mosses, quillworts, horsetails, and whisk ferns, (pteridophytes are now categorized as Lycophytes and Euphyllophytes (after Smith et al., 2006). Once considered of the same evolutionary line, all these plants were formerly placed in the single group Pteridophyta and were subsequently referred to as the ferns and *fern* allies. Although modern studies have shown that these plants are not in fact related, yet these terms are still used in discussion of the lower vascular plants. Pteridophyte Phylogeny Group or PPG, is an informal international group of systematic botanists who have collaborated to establish a consensus on the classification of pteridophytes (lycophytes and ferns) based on the knowledge

---

Vascular plants are those that possess a specialized conducting system for the transport of water, minerals, and food materials, as opposed to the more primitive *bryophytes*—mosses and liverworts—which lack such a system. They include lower vascular plants which lack vessels in xylem and that reproduce by *spores* and both the seed plants—*angiosperms* and *gymnosperms*, with common occurrence of vessels in the former.

about plant relationships discovered through phylogenetic studies. The PPG 1 recognizes two unrelated classes of extant lower vascular plants: Polypodiopsida, the ferns and horsetails, and Lycopodiopsida, the lycophytes.

The lower vascular plants represent the oldest of land plants (Fig. 1.1). The early evolution occurred during the Devonian and Carboniferous periods, from 419.2 million to 298.9 million years ago. The conquest of land must have been a long hazardous and costly process of slow adjustment to a new and inhospitable environment. Many forms must have appeared and became extinct in the process while many others must have evolved to survive the new habitats. The sphenophytes, for example, were once a large and diverse group of herbs, shrubs, vines, and trees but are now limited to only 15 species of horsetails; while the woody lycophytes (club mosses) are entirely gone, leaving only a faint trail in their reduced modern representatives. On one hand is the fossil fern foliage of the Carboniferous Period of seed ferns such as Archaeosperma, which are the probable antecedents of the flowering plants while on the other hand are modern ferns that represent an explosion of evolution in Cretaceous times (145 million to 66 million years ago) (See Box 1).

Box 1: Fern explosion

In today's angiosperm-dominated terrestrial ecosystems, leptosporangiate ferns are truly exceptional - accounting for 80% of the 11,000 non-flowering vascular plant species. This remarkable diversity according to recent research is mostly the result of a major leptosporangiate radiation that began in the Cretaceous, following the rise of angiosperms.

Though ancient, it appears that ferns really took to luxuriant growth during a very hot, very wet period that peaked about 10 million years after the Cretaceous/Tertiary boundary 65 million years ago. It must be remembered that there was also a rise of angiosperms during the Cretaceous period and ferns are said to have diversified in shadow of angiosperms.

Two main innovations led to the ferns' success in the face of the new competition from flowering plants which appeared on the scene. One was the ability to make a living on light that was more toward the red end of spectrum -- shade, in other words. And second was that some ferns also developed the ability to live on trees, sometimes without soil, as epiphytes. By storing water, developing thicker skin that would cut down transpiration, or being more tolerant to drying out, the epiphytic ferns could now climb on a trunk, or twig and live quite happily more than 100 feet off the forest floor, where moisture, temperature, and sunlight were very different indeed. So, as rain forests developed and tropical trees and vines clawed past each other to reach heights, they took the ferns up along with them. Thousands of new fern species evolved to take advantage of all the new niches being created in the canopy which also meant escaping the battle of survival on the ground!

Pteridophytes are not considered an economically important group although they are used by local communities around the world for medicines and food. Their greatest value today is in horticulture (ferns) and their remains provide the bulk of the world's coal beds. Furthermore, their relatively simple structure and life cycle makes them extremely valuable to researchers in understanding the overall process of plant structure and evolution.

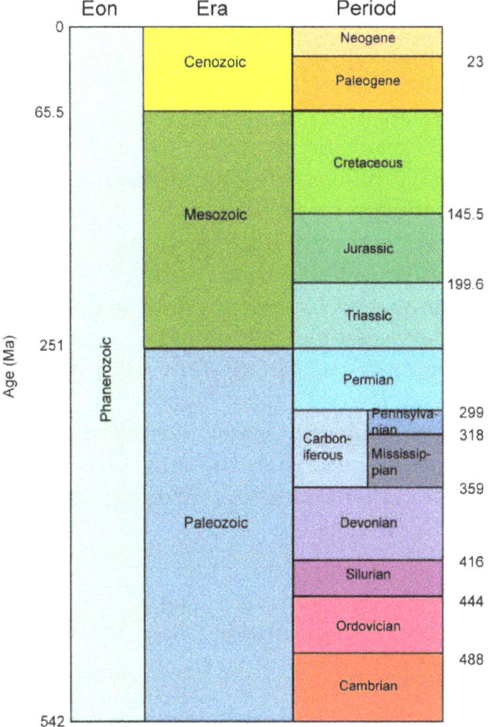

Figure 1.1: Geologic time scale

VASCULAR SYSTEM – BEGINNING OF TERRESTRIAL LIFE

The conducting system of lower vascular plants includes the xylem, composed largely of tracheids (tubular cells) for transport of water and minerals; and the phloem (sieve cells) for translocation of food materials. These vascular tissues are arranged in different patterns in different plant groups and in different parts of the plant. In pteridophytes, a whole range of stelar evolution is encountered.

The vascular cylinder of a stem or root is called the stele. The simplest and apparently most primitive type of stele is the protostele, in which the xylem is in the centre of the stem, surrounded by a narrow band of phloem. It is bounded by a pericycle composed of one or two cell layers and a single cell layer of endodermis. While pericycle is generally the layer giving rise to the branches in roots, the endodermis seems to regulate the flow of water and dissolved substances, from the surrounding cortex. More common in fern stems are siphonosteles, having a pith in the centre with the vascular tissue forming a cylinder around it. In megaphyllous fern, leaf is attached to a stem, and a part of the vascular tissue of the stem goes into it (a leaf trace), creating a slight gap, filled by parenchyma cells (generalized plant cells), in the vascular cylinder. If the leaves are distant and the stem long and creeping, a single gap will be seen in cross section; if leaves are close together or numerous, the gaps overlap, and affect the continuity of the stele, causing the cylinder to appear in cross section as a ring of disconnected round or elongate bars (meristeles) of vascular tissue.

Generally in extant pteridophytes, when the young organs mature, no further growth in diameter takes place. In several extinct groups a special ring of cells, the cambium, produces additional xylem (secondary xylem) to the inside and phloem cells (secondary phloem) to the outside (secondary growth as opposed to primary growth achieved by apical activity of the stem and root), resulting in increased diameter and a truly woody plant. Secondary growth by means of a vascular cambium has repeatedly arisen during the evolution of vascular plants. Plants like *Lepidodendron* and *Calamites* showed conspicuous secondary growth. This is common in many seed plants today, but in the extant lower vascular plants only two genera (*Botrychium* and *Isoetes*) show a slight vestige of secondary growth. Even in present day tree ferns (*Cyathea*, *Dicksonia*, *Cibotium*), with trunks up to 25 metres (80 feet) tall, the tissues are entirely the result of growth from the stem apex and not from secondary tissues. Their strength is derived not from woody growth in diameter but by strengthening tissues surrounding the vascular bundles and in some cases by a mantle of roots.

The cells of the vascular strands in pteridophytes are mainly tracheids, sieve cells, parenchyma, and endodermal cells. The tracheids, which comprise xylem, or water-conducting tissue, are normally long, narrow, and attenuated at the tips. Their secondary walls show scalariform thickenings. The largest tracheids are several centimetres long, but more commonly they are much smaller. Vessel cells, which have evolved in several lines of fern evolution and are the principal water-conducting cell type of flowering plants, provide direct, unimpeded connections for water transport between the cells. Vessels, longitudinal channels composed of linear series of such perforated cells, have been reported from such diverse ferns as waterclover (*Marsilea*), bracken (*Pteridium*) and also from certain species of *Selaginella*.

The phloem is composed mainly of sieve cells - narrow, elongated units with persistent protoplasts and nuclei (i.e., they are still alive at functional maturity) and lack secondary walls with elaborate pitting as seen in xylary elements. Sieve cells usually display more or less distinguishable sieve-like areas, through which, presumably, organic food is translocated to the stem and other plant organs. There are various arrangements of xylem and phloem in the vascular system, but usually a single strand composed of both is surrounded by parenchyma cells, the pericycle, and an outer layer of cells with specialized walls, the endodermis. Endodermal cells in young stems are provided with special strips of secondary wall material known as casparian strips on their radial walls. As the stems age, there is a tendency for the endodermal cells to become thick-walled around the entire circumference. The pith is made up of parenchyma cells as a rule, but, in some fern genera, scattered tracheid-like cells are found as well.

In the vascular tissues, the direction of radial maturation of successive procambial cells that affects the differentiation of the primary xylem is important. In the roots of all vascular plants and stems of *Psilotum* and the Lycopsida, the first protoxylem cells to acquire secondary walls occur at the outermost edge of the procambial cylinder. These cells establish the future pattern of xylem differentiation which occurs centripetally or toward the centre of the axis. Such a primary xylem is termed exarch and is regarded as the most primitive condition in vascular plants. In stems of modern seed plants, protoxylem begins its development from the inner most procambial cells, that are situated next to the pith and the remainder of primary xylem differentiation occurs centrifugally or toward the periphery of the stem. This type of primary xylem is termed endarch xylem. In the leaf and stem bundles of many ferns, the primary xylem is mesarch. It is a condition where the protoxylem begins development within the procambial strand and further xylem formation occurs both centripetally and centrifugally; consequently at maturity, the protoxylem cells are surrounded by the metaxylem.

SPOROPHYTE

The sporophyte is the dominant phase in life cycle and it is much differentiated (with roots, stem and leaves) when compared to that of bryophytes. Taproots are unknown in lower vascular plants. Embryonic root in pteridophytes is short lived and roots on adult plants are adventitious structures as they arise at points along the stem. In construction, the roots are generally regarded as being much less diverse than the stems. They are protostelic, lacking pith and gaps, and they grow from one or more apical initials (cells that divide to produce all the cells and tissues of an organ). The root hairs have a role in absorption of water and nutrients and in attachment of the plant to the soil or other growing surfaces. The endodermis of the root is well marked, and casparian strips are present, as in the stem. It becomes thick-walled and hardened (sclerified) at maturity. The production and development of xylem tissue in the steles of most pteridophyte roots is diarch.

Stem appendages known as leaves take various forms in different groups of pteridophytes. The simplest are scale-like emergences, or enations, that are not supplied by vascular tissue (i.e., they have no veins), found in some extinct groups and in modern whisk ferns (*Psilotum*). The lycophytes have scale-like, needle-like, or awl-shaped "microphylls" with a single, unbranched vein. The sphenophytes have "sphenophylls"—scale-like leaves with a single vein in the modern *Equisetum* or wedge-shaped leaves with a dichotomously forking vein system in many of the fossil forms. These leaf forms are all small and simple and the vascular connection with the stem stele does not affect the stele configuration, causing no leaf gap. On the other hand, the complex leaves of ferns (pteridophylls, or megaphylls) which have probably evolved from a branching stem system, affect the stele by drawing out enough vascular tissue to cause a leaf gap. Thus, in ferns the presence of megaphylls is correlated with some type of siphonostele, i.e., a central pith is enclosed by a continuous or dissected cylinder of vascular tissue.

REPRODUCTION THROUGH SPORES

The pteridophyta are polysporangiate, in contrast to bryophytes which bear one sporogonium per sporophyte. In *Horneophyton*, *Rhynia* and *Cooksonia,* the fossil forms, sporangia were cauline, while in most of the extant species sporangia are associated with leaves qualifying them as sporophylls. In lower pteridophytes, the sporophylls are aggregated to form compact cone-like structures, strobili. The sporangia on the basis of their development are classified into two types – eusporangiate and leptosporangiate. In the former type, characteristic of lower pteridophytes and some ferns, the sporangium originates from a group of cells. In the latter, the sporangium originates from a single initial and is characteristic of higher ferns also named leptosporangiate ferns. Bower reported the occurrence of transitional forms in Osmundaceae.

The sporangial initials in the eusporangial type are superficial in position and periclinal divisions of these cells result in outer wall cells and inner primary sporogenous cells. e.g. *Psilotum, Tmesipteris, Lycopodium, Selaginella, Isoetes, Equisetum* and eusporangiate ferns. The leptosporangium originates from a single superficial cell and the mature sporangium has a wall consisting of a single layer of cells.

EVOLUTION OF SORUS IN FERNS

In ferns, sporophylls are similar to foliage leaves and are photosynthetic organs. As a departure from this generalization, is the production of sporangia in restricted regions of the sporophylls. As a result, some portions become strictly reproductive and non-photosynthetic. In an extreme case, the entire leaf produces only sporangia. This results in dimorphic condition on the stem, i.e., sterile and fertile leaves. In certain forms like *Marselia,* the sporangia are produced in specialized structures, the sporocarps. The sporangia in higher ferns are aggregated in groups known as sori. If sporangia do not form sori, they are scattered over the lower leaf surface forming a felt or marginal tassels over the surface of slender sporophyll (fertile leaf) divisions as in *Osmunda* or are solitary along or close to the margins of narrow leaf segments as in *Schizaea* and *Anemia.*

On the basis of their origin, the sori are classified into the following types:

a. Marginal sorus: In this type, the receptacle and sporangia originate from the margin of pinnae or pinnule. The indusium if present is a submarginal outgrowth around the receptacle and is funnel shaped or two lipped.

b. Intramarginal sorus: Sori of many ferns appear to be marginal but when closely observed are submarginal in origin. The marginal cells often form a thin extension or exhibit loose meristematic activity. The submarginal cells on abaxial side of the lamina become meristematic forming the receptacle or sporogenous meristem. In some instances a thin abaxial indusial flap is also formed from the surface cells near the meristem. eg., *Pteris* and *Pteridium.*

c. Superficial or Abaxial sorus: In such types, the receptacle is submarginal in origin on lower surface but the margin of the lamina remains active and continues to add new tissue shifting the sorus farther away from margin.

The indusium if present, is formed from the superficial cells of receptacle and overarches the sources.

There may however, occur situations which are not clearly demarcated. In certain situations, during ontogeny the sori of marginal origin may be shifted towards abaxial surface and sori of abaxial origin at maturity may appear to be marginal due to differential growth of sorus and surrounding tissue.

The protection to young sporangium during ontogeny is provided by circinate vernation. At maturity, however in the expanded sporophyll, the sporangia are with or without a protective covering, the indusium. The sori lacking indusium are exindusiate or naked. In forms having, marginal sori, the indusium might become a cup or pouch-like structure formed by extension of the abaxial and adaxial margins of the sporophyll (*Duvallia)*. In forms having superficial sori, the indusium is essentially an outgrowth of sporphyll (leaf epidermis). When present, indusium is variable in its form and extent. It may be a delicate strip attached along one side only (unilateral *Asplenium*), reniform and attached at sinus (*Dryopteris*), a peltate disc (*Polystichum*), circular, horse-shoe shaped or cup-shaped. Instances where reflexed leaf margin provides protection to sporangia, the structure is termed as false indusium.

GAMETOPHYTE

The plant begins life as a haploid spore. The germinating spore grows into a small gametophyte, or prothallium, usually only 0.3 to 1 centimetre (0.2 to 0.4 inch) long or broad, bearing rhizoids (hair-like structures for water and mineral absorption and attachment to the soil). Gametophytes may be green, occurring on the soil surface, or colourless, occurring under the soil (usually saprophytically, acquiring nutrients with the aid of a mycorrhizal fungus). Sex organs, called antheridia and archegonia, that produce antherozoids and eggs, respectively are borne on the gametophyte. The antherozoids require water to swim to the egg for fertilization. Thus, like bryophytes, pteridophytes depend on water for fertilization. The zygote then divides, developing into an embryo, which in turn develops the first leaf, root, and stem apex. The resulting plant, the sporophyte at maturity produces sporangia (spore cases); in them the spore mother cells divide by meiosis resulting in four haploid spores.

In majority of the pteridophytes all the spores are similar, making these plants homosporous. A few groups (the lycophytes - *Selaginella* and *Isoetes* and, among the ferns, the water-fern families Marsileaceae, Salviniaceae, and Azollaceae) produce two kinds of spores - mega/macro (large) and micro (small) spores, making them heterosporous. The megaspore divides to form a mass of tissue and archegonia, enclosing a potential female gamete - the egg. The megaspores hold food reserves for the early development of the embryo. On the other hand, microspores divide to form a reduced gametophyte comprising merely a jacket of cells and a few potential antherozoid mother cells.

In the homosporous type, the gametophyte is exosporic i.e., free living and not enclosed by the spore wall whereas in all heterosporous groups the micro and megagametophytes are endosporic i.e., entirely or for the most part enclosed by the wall of the microspore or megaspore respectively. Exosporic gametophytes are either free living or nourished through mycorrhizal association. They are thalloid with rhizoids as organs for attachment and are commonly referred to as prothallus (singular), which in *Schizaea* is protonemal. In *Equisetum*, the prothallus shows certain differentiation of parts while in some species of *Lycopodium*, the lobes of the prothallus are leaf-like. The prothalli are typically monoecious with potential to produce both types of gametangia. However, in some instances, the environmental conditions surrounding the developing gametophyte are decisive with respect to the gametangium production. Unfavourable growing conditions such as those commonly resulting from crowded masses of gametophytes, may cause inhibition of archegonia. This is true of prothalli of *Equisetum* and some of the leptosporangiate ferns where weak or impoverished gametophytes form only antheridia. The antheridia may entirely or in part remain sunk in gametophytic tissue – the embedded type, or they may project from it – the emergent type, usually found in the leptosporangiate ferns. The gametangia of exosporic gametophytes are usually embedded in the vegetative tissue of the prothallus. This is always the case with the venter or egg containing portion of the archegonium, but the antheridium is less consistent. In the homosporous members of the Lycopsida, in *Equisetum* and in the eusporangiate ferns, the sterile jacket of the antheridium protrudes slightly above the surface and the spermatogenous tissue is deeply sunken. But in Psilotales and the leptosporangiate ferns, the antheridia are conspicuously the emergent organs.

The distribution pattern or the position of antheridia and archegonia varies considerably and is correlated to some degree with the form and level of evolutionary development of the gametophyte. In the fleshy, radial non-photosynthetic gametophytes of *Psilotum* and *Tmesipteris,* the two kinds of gametangia are intermingled and tend to occur over the entire surface of the gametophyte. An intermingling of sex organs also occurs in certain presumably primitive species of *Lycopodium* while in others antheridia and archegonia are formed in distinct patches or clusters on the upper surface of the gametophyte. The antheridia develop first and are borne centrally and the archegonia appear later and are positioned near the margins. In thalloid, dorsiventral photosynthetic type of gametophyte characteristic of the Marattiaceae and leptosporangiate ferns, several somewhat intergrading patterns of sex organ distribution occurs. In higher groups of leptosporangiate ferns, both kinds of sex organs are commonly restricted to the lower surface - the archegonia are limited to the cushion of tissue situated behind the notch of the heart shaped prothallus, while the more numerous antheridia occur near the basal end of the prothallus as also on the wings.

Endosporic gametophytes are usually dioecious. Male gametophyte is extremely reduced in structure consisting as in *Selaginella* and *Isoetes* of a single vegetative prothallial cell and one antheridium. The megagametophyte by contrast is more robust consisting of a mass of food storing tissue which usually fills the cavity of the megaspore and which at least in *Selaginella* is exposed by the cracking of the spore wall along the triradiate ridge. In some ferns, as in *Marselia*, even the megagametophyte shows extreme reduction and consists of single archegonium and a few vegetative cells. The small archegonia are restricted in occurrence to the surface of the protuberant cushion of gametophyte tissue below the cracked region of the spore wall. The female gametophytes are retained on the parent (sporophytes) for variable periods. Post fertilization, the development of the zygotes into embryos takes place within the female gametophytes. This retention of the fertilized product on the parent is an important event in evolution that has led to the seed habit prevalent in higher plants.

EMBRYOGENY

The polarity of the embryo in *Lycopodium, Selaginella, Isoetes,* eusporangiate ferns is described as endoscopic. The less frequent exoscopic polarity characterises the embryogeny of *Psilotum, Tmesipteris, Equisetum* and some members of Ophioglossaceae. The presence of suspensor is invariably correlated with an endoscopic type of polarity. Following the early definition of polarity, the establishment of an eight celled embryo, the young sporophyte enters upon a phase of enlargement and organ formation.

According to Bower (1922), a very young embryo in pteridophytes is somewhat filamentous or spindle shaped and consists of two definable poles: the apical pole which gives rise to the terminal meristem or shoot apex and one or two leaves and the basal pole represented in many plants as suspensor. The first leaves are designated as cotyledons, which may be very small or rudimentary as in *Lycopodium* while in *Isoetes* and many leptosporangiate ferns they are prominent and photosynthetic. A wide variation in suspensor formation is seen. In *Lycopodium* and *Selaginella* embryos have suspensor, while *Isoetes* does not have one. In *Psilotum* and *Tmesipteris* embryo is without root, while in other pteridophytes roots are strictly lateral organs. The embryo of these lower land plants also show foot formation. Bower (1935) regards foot as an opportunistic growth which arises in position convenient for performing its function as a sectorial or nursing organ. In some embryos such as those of *Selaginella, Isoetes, Equisetum* and *Ophioglossum* and in leptosporangiate ferns, the foot is conspicuous and its role in anchoring and conveying nutrients from the gametophyte is evident. Perhaps the most impressive example of sectorial foot is found in the embryos of *Tmesipteris* and *Psilotum* where the entire lower half of the embryo develops into a haustorial structure. This structure sends lobes and irregular processes deep into the tissue of the gametophyte.

ALTERNATION OF GENERATIONS

The life cycle of pteridophytes exhibits an *alternation of generations* between gametophytes and sporophytes. The gametophytes are sexual plants producing eggs or antherozoids (dioecious) or both (monoecious), and the sporophytes are asexual plants bearing spores that are capable of producing new gametophytes. The sporophyte of lower vascular plants, in contrast to that of mosses and liverworts, is the dominant generation capable of photosynthesis and therefore, independent phase. The life cycle of the lower vascular plants is basically the same as that of seed plants. The main difference is that in seed plants the new young sporophyte (embryo) is raised within a structure (seed) on the parent plant before dispersal and passes through a resting stage, whereas in lower vascular (pteridophytes) plants dispersal and resting takes place in the spore before the embryo develops further.

Table 1.1: Comparison between bryophytes and pteridophytes

Character	Bryophytes	Pteridophytes
Body plan	Leafy or thalloid	Plant body differentiated into roots, stem and leaves
Vasculature	No xylem and phloem	These tissues are present
Roots	Absent	Present
Archegonium	Exposed with six rows of cells	Generally embedded with four rows of cells
Antheridium	Stalked	Sessile
Dominant phase	Gametophyte while sporophyte is dependent	Sporophyte is independent and dominant
Prothallus	Spore germinates to form a germ tube in liverworts and protonema in mosses	Spore germinates to form prothallus which shows a wide range of structure
Apogamy/apospory	Not frequent	Is common
Embryogeny	Embryo forms two fundamental tissues – endothecium and amphi-thecium	Embryo is a much elaborate structure with well-defined polarity
Spores	The entire produce of the sporo-gonium is similar	Spores may be similar while in some forms heterosporous condition prevails
Sporophyte	Does not show true leaves and stem	The sporophyte has true leaves , stem and roots
Economic impor-tance	A large number of species are of economic and ecological impor-tance	The plants have limited importance and are more common in horticulture

SALIENT FEATURES

- Pteridophytes enjoy a position between bryophytes and spermatophytes.
- They resemble bryophytes and gymnosperms in possessing archegonium, a structure which has pooled all the three groups under Archegoniates.
- The male sex organ is antheridium in which are produced motile antherozoids.
- Fertilization is still dependent on water and together with bryophytes, pteridophytes are occasionally referred to as 'amphibians of plant kingdom' (Table 1.1.)
- They show distinct alternation of generations. The sporophytic phase ends with the production of non-sexual spores.
- However, the sporophyte is independent, dominant, photosynthetic, and organized into roots, stem and leaves.
- Pteridophytes have evolved to fill up every ecological niche on earth but greatest species variety is seen in the tropical rain forests.

Many species of pteridophytes are disappearing due to degraded ecosystems and to document the fern flora and its status, there is the International Association of Pteridologists (IAP). As the fern species started getting endangered, IAP appointed a committee for fern-conservation. This became the Species Survival Commission (SSC) specialist group for Pteridophytes, which, using criteria established by the IUCN, developed a list of threatened ferns world-wide. Periodic surveys are required to update such lists and conservation strategies can be developed based on this data.

THE FOSSIL PTERIDOPHYTES AND EVOLUTIONARY LINK

In 1859, Canadian geologist Dawson described, the first fossils of vascular plants from Devonian peninsula of Canada. The reconstruction of this plant revealed slender axes that arose from a subterranean rhizome system which bore rhizoids. The aerial axes were dichotomously branched, had scale-like appendages and terminated into sporangia. Because of its resemblance to *Psilotum,* Dawson called it *Psilophyton princeps.* It was however, in 1917 when Kidston and Lang described similar plants from Lower Devonian red sandstone cherts of village Rhynie. The fossil plants were grouped under Psilophytales. The order includes a group of ancient vascular plants with great importance to the morphology and phylogeny of higher plants. Interest in these ancient plants lies in their great age and simplicity of structure - they are the oldest known land plants of upper Silurian and Lower and Middle Devonian age. These plant forms suggest the beginning of a vascular land flora and provide information as to the origin of land pants. *Rhynia, Psilophyton, Asteroxylon, Hornea, Pseudosporochnus, Drepanophycus* and *Baragwanathia* are some of these ancient vascular plants (see Fig. 1.1).

Three genera –*Rhynia, Hornea* and *Asteroxylon* have been described from Scotland.

Rhynia

Is the simplest genus, with two species —*R. major* and *R. gwynne-vaughani* with a slender rhizome from which arose an erect cylindrical dichotomously branched stem. The plants were anchored by rhizoids that arose from rhizome, roots were absent. The stems were with blunt apices and were naked or without any leaves. The stems of *Rhynia major*, the larger species were about 50cms in height and 1.5 to 6mm in diameter; those of *R. gwynne-vaughani* were 2 or 3mm in diameter and upto 20cm in height. The latter species bore on its smooth stems, hemispherical or oval protuberances, the nature of which remains unclear. These projections were mere parenchymatous bulges or cushions, somewhat lenticel-like in form and structure. When situated at the base of the stem, they apparently in some cases bore rhizoids. Some seem to have grown out into small lateral branches which were easily detached, indicating a possible means of vegetative reproduction.

Anatomy – The axis had a slender central protostele with a cylindrical strand of xylem surrounded by a sheath of elongate cells with oblique ends comprising phloem. The cortex was thick while the delicate cutinized epidermis had stomata similar to those of other vascular plants. The tracheids were all annular and the central ones were smaller. Intervening pericycle and endodermis was not clear.

Reproduction – The sporangia were cylindrical structures borne terminally on the branch tips. They were continuous with the stem and had a slight constriction at the base forming an indefinite short stalk. Those of *Rhynia major* were large about 12mm long by 4mm in diameter. The sporangial walls were massive formed of several layers of cells with the outer ones thick walled. No particular means of dehiscence was seen. The many spores, all alike and cutinized were borne in tetrads and filled the sporangium chamber. The inner wall cells probably formed an indefinite tapetum.

In their description of *R. gwynne-vaughani* Kidston and Lang (1917) did not mention organic connections between sporangia and axes. Pant (1962) suggested that the axes were more like a gametophyte than a sporophyte and the spherical projections on the axes were possibly young sporophytes arising from the gametophyte. Lemoigne (1968) showed that these structures were archegonia-like. It is therefore, more likely that *R. gwynne-vaughani* is the gametophytic generation of *R. major*.

Hornea

Was similar to *Rhynia* but was smaller and a more slender plant with somewhat more freely branched stem which was without appendages of any sort. The rhizomes were shorter and jointed with thick tuberous segments.

Anatomy – The stele of the stem was similar to that of *Rhynia*, but the rhizome was apparently without vascular tissue, consisting wholly of parenchymatous cells. The steles of the attached aerial stems passed into the fleshy mass, expanded bell like and faded out. These rhizome joints have been compared to protocorms in their anatomy and in their relation to the vascular axes.

Reproduction – The small sporangia were terminal on the stem tips and sessile, being continuous with the stem. They had no uniform shape, appeared merely as fertile stem tips and in many cases they forked dichotomously, like the stem. They possessed a thick wall and showed no evidence of dehiscence. The spores lay in a dome shaped cavity around and above a central columella, which presumably represented a continuation of the phloem of the stele.

Asteroxylon

Two species are well known in this genus: *Asteroxylon mackiei* from Scotland and *A. elberfeldense* from Germany. The former was discovered first.

The plant body was much more complex than that of *Rhynia* and *Hornea*. A naked branching rhizome sent up freely branching erect stems that bore small lycopodium-like leaves. At the base of the leafy axes there was a transition to a naked region below. The more slender underground branches of the rhizome were root-like in appearance and behaviour, some of them burrowing into the decaying tissues of other plants.

The leaves were about 5mm in length and closely crowded, apparently without regular arrangement. In form they were more spine-like, being conical or awl shaped, in cross section appearing oval or round-triangular.

Anatomy – The vascular cylinder was more complex than that of *Rhynia* and *Hornea*. The stele in cross section in the smaller species was stelate protostele, while in the larger species, a polystelate stele with a central pith was seen. From the arms of the stele small vascular strands extended obliquely upward to the bases of the leaves, which were veinless. The tracheids of the smaller species were like those of *Rhynia*, while those of the larger included scalariform types and some had more or less rounded pits. The smallest cells were near the outside leading to a mesarch or exarch condition. There was no secondary growth.

Reproduction –No sporangia were at first found in connection with plants, but associated with the leafy stems were slender naked dichotomous branches bearing terminal sporangia. These naked branches owing to their method of occurrence and to the similarity of their stomata and other anatomical structure, were suspected of being the deciduous fertile branchlets. In *Asteroxylon elberfeldense*, occasional connections of the two types of stems were found and the transition is thought to have followed a pattern of leafy branches that passed into spiny ones and those into naked fertile tips. The sporangia were ovoid or ellipsoid, and minute with a simple annulus and apical longitudinal dehiscence.

ARE RHYNIALES THE OLDEST VASCULAR PLANTS?

Leclerecq (1954) remarked that Devonian Rhynie flora in general and *Rhynia* in particular is by no means the only representative of the earliest land plants. More complex forms – *Asteroxylon* coexisted with *Rhynia* and more complex forms such as *Aldanophyton*, preceded them in fossil record. Therefore primitive status of *Rhynia* is questionable. Nevertheless, it is possible that *Rhynia* attained its present structure and did not evolve further but existed in the form for millions of years. In that case it is truly primitive. Alternatively, if *Rhynia* is a reduced form arising from more complex land plant, then it is not a primitive plant. More significant about *Rhynia* is its simple nature which lacked differentiation into stem and roots, and its appendages could not be called leaves. Its stele was a slender strand of tracheids while the sporangia were modified branch tips. Therefore irrespective of whether *Rhynia* is a reduced form from complex ancestors or is truly a primitive form, in its simple organization it represents one of the stages in evolution of vascular land and plants.

Vascular plants have also been discovered from Silurian period. In 1935, Cookson and Lang described a few plants from Silurian of Australia. *Cooksonia* from Silurian of Britain had naked dichotomously branched axes with a thin vascular strand. The axes terminated into sporangia nearly spherical in *C. hemispherica* and wider than long in *C. pertonii*.

Yarravia, another ancient plant from Silurian of Australia had naked axes which terminated into large synangia. Each synangium comprised a large number of elongate sporangia.

The best known is *Zosterophyllum*. It had dichotomously branched vegetative system from which arose fertile branches terminating into spike-like structures bearing sporangia. *Z. rheananum* differs from *Z. australianum* in having flattened stem and is suggested to be aquatic.

Trends in evolutionary speciation:

Kidston and Lang (1917) gave following lines of evolution

a. *Rhynia, Horneophyton* and *Cooksonia* represent the first line with terminal sporangia that dehisced longitudinally.

b. *Zosterophyllum* represents the second line in which sporangia were lateral, globose or reniform and dehisced along distal edge.

c. *Trimerophyton* along with *Psilophyton* is the third line. In *Trimerophyton* sporangia terminate dichotomously or trichotomously branched lateral axes.

QUESTIONS

Q1. Briefly discuss the salient features of pteridophytes.

Q2. Why pteridophytes are also called reptiles of plant kingdom?

Q3. What enabled pteridophytes grow luxuriantly in face of angiosperm invasion?

Q4. What importance does *Rhynia* have as a link between Bryophytes and Pteridophytes?

2

CHAPTER

CLASSIFICATION

Pteridophytes (Lat.) or (Lat. Pteridophyta) in the broad interpretation of the term, are vascular plants that reproduce via spores. They are better interpreted as cryptogams because of the absence of flowers and seeds. The group includes ferns, horsetails, club mosses and whisk ferns that do not form a monophyletic group. Therefore, pteridophytes are no longer considered to form a valid taxon, but the term is still used as an informal way for the sake of convenience, to refer to the above mentioned plants.

In some early classifications proposed by Smith (1955), Bold (1957) and Zimmerman (1959), pteridophytes were divided into Psilophyta, Lycophyta, Sphenophyta and Pterophyta. In commemoration of the 250th birthday of Carolus Linnaeus, Pichi-Sermolli (1958) provided an overview of higher classification of pteridophyta which embraced ferns and 'fern allies'. According to him, classifications until then had reflected four trends; (1) four major lineages usually treated as classes were accepted: Psilopsida, Lycopsida, Sphenopsida and Pteropsida, the last including ferns and seed plants ; (2) ferns and 'fern allies' were treated as a group but Psilopsida was excluded, treating the latter as related to Bryophyta or as an independent lineage together with the fossil Psilophytopsida; (3) maintaining Pteridophyta in their classical sense, including all vascular cryptogams (e.g. Campbell, 1940; Reimers, 1954); and (4) recognizing the independent divisions Psilophyta, Lycophyta, Sphenophyta, Pteridophyta and Spermatophyta (Benson, 1957). Pichi-Sermolli effectively dismissed trends 1, 2 and 4, concluding that all pteridophytes should be treated in a single division in the classical sense, and placed them next to Spermatophyta.

HISTORICAL OVERVIEW

As with all botanical studies, the nomenclatural history of fern taxonomy started with Linnaeus (1753, 1754), who first attempted to organize ferns based on the shape and position of sori on leaves. This use of a single character produced an artificial but easily used classification. Fern diversity and anatomy had been addressed in pre Linnaeus period as well (e.g. Cesalpino, 1583; Malpighi, 1675; Grew, 1682; Morison, 1699; Van Rheede tot Drakestein, 1678–1703; Plumier, 1705; Petiver, 1712), but it remained a Herculean task as little was known about the life cycle of ferns. The sexual system of Linnaeus, based on the number of stamens and pistils (the so-called 'sex organs' in flowering plants), worked well for artificially classifying angiosperms, but it was impossible to place ferns in a system based on numbers of male and female parts. Ferns were therefore, placed in Cryptogamae, among mosses, algae, fungi and even some animals (corals, sponges). In his Cryptogamia, Filices, Linnaeus (1753, Sp. Pl. 2: 1085–1100) recognized 16 genera and 174 species: *Equisetum* L. (6), *Onoclea* L. (1), *Ophioglossum* L. (6), *Osmunda* L. (17), *Acrostichum* L. (25), *Pteris* L. (19), *Blechnum* L. (2), *Hemionitis* L. (2), *Lonchitis* L. (3), *Asplenium* L. (20), *Polypodium* L. (58), *Adiantum* L. (15), *Trichomanes* L. (11), *Marsilea* L. (2), *Pilularia* L. (1) and *Isoe¨tes* L. (1). This grouping is still valid excepting for *Isoetes* which is classified as a lycopod and does not have a close relationship to the other genera. Linnaeus's classification was artificial, and species of clearly distant relationships were placed together. For these simple reasons, Linnaeus's classification was not convincing in general, but it served the purpose of organizing the numerous newly discovered species until a more 'natural' system was made available. In the meantime, Swartz (1806) tried to classify ferns and in his book - Synopsis Filicum and listed 33 genera and about 700 species.

Ferns remained biologically poorly understood and occasionally the authors may not have known in which genus to place a new species and thus a guess must have been made, which was subsequently corrected in some new system of classification. Then came Hooker and Bauer's (1842) system of classification where it was emphasized that characters of 'fruiting parts' were more important than characters of 'vegetative parts' for classifying all plants at the generic level. However, even this character worked better with angiosperms as the concept of family was long established in flowering plants with many families having a long history of recognition. Unfortunately, the emphasis on 'fruiting parts' for ferns did not result in either a consistent or a widely accepted system. Hence, there was re-application of the Swartzian system, wherein characters of sori and indusia were emphasized and which lasted for another 50 years as the most often used fern classification. Reimers (1954) provided a developmental scheme in which fern lineages were set out on a time scale, all essentially originating from Psilophytales (in which he included the fossils *Rhynia* and *Psilophyton*) in the early Devonian (320 million years ago). The attempts to classify these plants (pteridophytes) remained for several decades in state of flux. The present classifications of Smith et al., (2006, 2008) and Christenhusz et al., (2011) have mostly reduced the number of genera, resulting in an expansion of several (e.g. *Asplenium, Blechnum, Cyathea* and *Hymenophyllum*). They have also resulted in the acceptance of narrower generic concepts in other groups (e.g. Hymenophyllaceae, Polypodiaceae and Pteridaceae). These classifications are, however, remarkable in that they combine morphological and molecular data to classify plants.

Box 2.1

With changing times, classification systems are now based on phylogeny. The Angiosperm Phylogeny Group, or APG, refers to an informal international group of systematic botanists who came together to establish a consensus view of the taxonomy of flowering plants (angiosperms) that would reflect new knowledge about their relationships based upon phylogenetic studies. The Angiosperm Phylogeny Group (APG) classification (APG, 1998; APG II, 2003; APG III, 2009) is not a complete formal classification of the angiosperms and recognizes only families and orders, with many major nodes unnamed or gives these only informal names e.g.,magnoliids, monocots, lamiids. This system was adopted by Chase and Reveal (2009) who thought it unwise to name these until the issues related to the ranking of these deeper nodes were resolved which according to them may happen in near future. It must be mentioned that many such taxonomic groups in the first APG classification (APG, 1998), which had many unclear relationships in the angiosperm tree, now stand resolved and a proper place has been assigned to these groups.

Working classification of class Equisetopsida [embryophytes] (after the land plants) Chase & Reveal (2009) :

Class Equisetopsida C.Agardh, Classes Pl.: 7. 20 May 1825

Subclass Anthocerotidae Engl., in H.G.A. Engler & K.A.E. Prantl, Nat. Pflanzenfam. 1, 3: 1, 6. 10 Oct 1893.

Subclass Bryidae Engl., Syllabus: 47.Apr 1892.

Subclass Marchantiidae Engl., in H.G.A. Engler & K.A.E. Prantl. Nat. Pflanzenfam. 1, 3: 1, 5. 10 Oct 1893.

Subclass Lycopodiidae Beketov, Kurs Bot. 1: 115. 1863.

[Monilophytes]

Subclass Equisetidae Warm., Osnov. Bot.: 221. 22-28 Apr 1883.

Subclass Marattiidae Klinge, Fl. Est-Liv-Churland 1: 93. 22-28 Jun 1882.

Subclass Ophioglossidae Klinge, Fl. Est-Liv-Churland 1: 94. 22-28 Jun 1882.

Subclass Polypodiidae Cronquist, Takht. & Zimmerm., Taxon 15: 133. Apr 1966.

Subclass Psilotidae Reveal, Phytologia 79: 70. 29 Apr 1996.

[Gymnosperms]

Subclass Ginkgooidae Engl., in H.G.A. Engler & K.A.E. Prantl, Nat. Planzenfam, Nacht. 1: 341. Dec 1897.

Subclass Cycadidae Pax, in K.A.E. Prantl, Lehrb. Bot., ed. 9: 203. 5 Apr 1894.

Subclass Pinidae Cronquist, Takht. & Zimmerm, Taxon 15: 134, Apr 1966.

Subclass Gnetidae Pax, in K.A.E. Prantl, Lehrb. Bot., ed. 9: 203. 5 Apr 1894.

[angiosperms]

Box 2.1 (Contd)

Subclass Mgnoliidae Novak ex Takht., Sist. Filog, Cvetk. Rast.: 51. 4 Feb 1967.

Superorder Amborellanae M.W. Chase & Reveal, stat. et superord. nov. Basionym: Amborellaceae Pichon, Bull. Mus. Hist. Nat. (Paris), ser. 2, 20: 384. 25 Oct 1948, nom. cons.

Superorder Nymphaeanae Thorne ex Reveal, Novon 2: 236, 13 Oct 1992.

Superorder Austrobaileyanae Doweld ex M.W. Chase & Reveal, stat et superord. nov. Basionym: Austrobaileyoideae Croizat, J. Arnold Arbor, 21: 404. 24 Jul 1940.

Under this APG III classification, class Equisetopsida has been created to include all embryophytes. **Monilophytes have been mentioned under class Equisetopsida with five subclasses which include the loosely constituted - pteridophytes.**

Earlier in 2004, Lewis & McCourt gave the classification of green plants in which major groups of green algae were recognized as classes (eleven) and stated that land plants should be given equal rank to the major groups of green algae, which they recognized as classes. It was argued that if Charales are placed under class Charophyceae, then all land plants (embryophytes) should be placed in a single class - Embryophyceae, leaving only the rank of subclass available for each of the major clades of land plants: mosses, hornworts, liverworts, lycophytes, whisk ferns, ophioglossoid ferns, marattioid ferns, osmundoid ferns and leptosporangiate ferns, cycads, conifers, *Ginkgo*, Gnetales and angiosperms. Chase & Reveal (2009) also discussed the possibility of recognizing the extant gymnosperms as a single subclass. However, this was not adopted because, by naming each of the gymnosperm groups as subclasses, it would have also to be extended to many of the fossil groups of gymnosperms, which are many, highly diverse and with unclear relationships to the extant gymnosperms. They also used taxonomic name Equisetopsida for all plants with embryo formation in contrast to Lewis & McCourt (2004) classification of green plants where the land plants were grouped under a class, Embryophyceae. But according to Chase & Reveal (2009) Embryophyceae was not a legitimate name because it was not based on a taxonomic name. Hence they retained the term Equisetopsida.

Thereafter, concepts of higher classification of ferns and lycopods have changed considerably throughout history. Fern allies have previously been loosely defined as vascular spore-bearing plants that are not ferns. They have been variously included in the lycopods (*Huperzia* Bernh., *Lycopodiella* Holub, *Lycopodium, Isoe¨tes* and *Selaginella*), horsetails (*Equisetum*), whisk ferns (*Psilotum* and *Tmesipteris*), Ophioglossaceae, water ferns (Marsileaceae, Salviniaceae and *Ceratopteris*) and sometimes even with members of Schizaeaceae and Gleicheniaceae. This was never considered a natural group but included genera that according to subsequent authors could not be placed among the ferns proper for various reasons. In most of the recent classification systems, this term is avoided because of its vague circumscription. In a similar way, the term 'eusporangiate fern' is to be avoided because this includes ferns that are not leptosporangiate; it only describes the plesiomorphic state (eusporangia are also present in seed plants and lycopods) and the taxa included do not form a clade. The water ferns (Salviniales) were previously placed among 'fern allies' because they are heterosporous like *Selaginella* and *Isoe¨tes.* They have however, been found to be sister to the polypods, so they are deeply embedded in the homosporous fern clade. Heterospory in general evolved several times independently in lycopods, ferns and seed plants, but this evolution is not necessarily the result of adaptation to an aquatic life form, although it appears as a plausible association for heterosporous ferns.

One of the major trends has been set by Chase & Reveal (2009) who consider Monilophytes under Equisetopsida with 5 sub classes (Box 2.1). Even though it is not a new concept (Linnaeus, 1753, already included *Equisetum* in his 'Cryptogamia filices'), the term 'monilophyte' was coined (Pryer et al., 2001a) for the fern clade including Equisetaceae and Psilotaceae, in contrast to the 'lycophytes', both of which are considered linguistically erroneous and superfluous terms. This clade has been traditionally called pteridophytes. However, some workers feel that the origin of the term 'monilophyte' is unclear, its etymology is obscure and has never been published as a formal taxon. 'Sphenophytes' (a term commonly used for extant horsetails and their fossil relatives) share a common ancestor with ferns, even though their exact placement within the fern lineage is still uncertain. It is therefore preferable to use Christenhusz & Chase (2014)-Trends and concepts in fern classification: Current consensus classification of extant Lycopodiophyta (lycopods) and Polypodiophyta (ferns). Estimated species and their numbers for some of the larger genera are also provided. In the linear sequence presented below, Christenhusz & Chase (2011) accept for lycophytes three families with five genera and for ferns, 45 families with about 280 genera. They have recognized each of the five major clades as subclasses, in line with the classification of Chase & Reveal (2009).

After Christenhusz et al., (2011) – All species are not listed.

[LYCOPODS; 3 families, 5 genera, approx. 1300 species]

SUB CLASS I LYCOPODIIDAE BEK.

ORDER A Lycopodiales DC.

FAMILY I Lycopodiaceae P.Beauv.1 (3/approx. 400). (*Huperzia* Bernh., *Lycopodiella* Holub, *Lycopodium* L.)

ORDER B Isoe¨tales Prantl

FAMILY 2 Isoe¨taceae Reichenb. (1/approx. 140) (*Isoe¨tes* L.)

ORDER C Selaginellales Prantl

FAMILY 3 SelaginellaceaeWillk. (1/approx. 750) (*Selaginella* P.Beauv.)

[Monilophytes]FERNS; 21 families, approx. 212 genera, approx. 10 535 species]

SUB CLASS II EQUISETIDAE WARM.

ORDER D Equisetales DC.

FAMILY 4 Equisetaceae Michx. (1/20) (*Equisetum* L.)

SUB CLASS III OPHIOGLOSSIDAE KLINGE

ORDER E Ophioglossales Link

FAMILY 5 Ophioglossaceae Martinov2 (4/approx. 80) (*Botrychium* Sw., *Helminthostachys* Kaulf., *Mankyua* B.Y.Sun, M.H.Kim & C.H.Kim, *Ophioglossum* L.)

ORDER F Psilotales Prantl

FAMILY 6 Psilotaceae J.W.Griff. & Henfr. (2/12) (*Psilotum* Sw., *Tmesipteris* Bernh.)

SUB CLASS IV MARATTIIDAE KLINGE

ORDER G Marattiales Link

FAMILY 7 Marattiaceae Kaulf.3 (6/approx. 130)

Subfamily Danaeoideae J.Williams (*Danaea* Sm.).

Subfamily Marattioideae C.Presl (*Angiopteris* Hoffm., *Christensenia* Maxon,

Eupodium J.Sm., *Marattia* Sw., *Ptisana* Murdock)

SUB CLASS V POLYPODIIDAE CRONQUIST, TAKHT. & ZIMMERM.

ORDER H Osmundales Link

FAMILY 8 Osmundaceae Martinov4 (4/approx. 25) (*Leptopteris* C.Presl, *Osmundastrum* C.Presl, *Osmunda* L., TodeaWilld.)

ORDER I Hymenophyllales A.B.Frank

FAMILY 9 Hymenophyllaceae Mart.5 (2/approx. 650) (*Hymenophyllum* Sm., *Trichomanes* L.)

ORDER J Gleicheniales Schimp.

FAMILY 10 Gleicheniaceae C.Presl (6/approx. 165) (*Dicranopteris* Bernh., *Diplopterygium* (Diels) Nakai, *Gleichenella* Ching, *Gleichenia* Sm., *Sticherus* C.Presl, *Stromatopteris* Mett.)

FAMILY 11 Dipteridaceae Seward & E.Dale (2/9) (*Cheiropleuria* C.Presl, *Dipteris* Reinw.)

FAMILY 12 Matoniaceae C.Presl (2/4) (*Matonia* R.Br., *Phanerosorus* Copel.)

ORDER K Schizaeales Schimp.

FAMILY 13 Lygodiaceae 1 genus *Lygodium* Sw.

FAMILY 14 Schizaeaceae Kaulf. (4/approx. 190)

Subfamily Schizaeoideae Lindl. (*Actinostachys* Wallich, *Schizaea* Sm.)

FAMILY 15 Anemiaceae

Subfamily Anemioideae C.Presl (*Anemia* Sw.)

ORDER L Salviniales Bartl.

FAMILY 16 Marsileaceae Mirb. (3/approx. 65) (*Marsilea* L., *Pilularia* L., *Regnellidium* Lindman)

FAMILY 17 Salviniaceae Martinov (2/approx. 20) (*Azolla* Lam., *Salvinia* Se´g.)

ORDER M Cyatheales A.B.Frank

FAMILY 18 Thyrsopteridaceae

Subfamily Thyrsopteridoideae B.K.Nayar (*Thyrsopteris* Kunze)

FAMILY 19 Loxsomataceae

Subfamily Loxsomatoideae Christenh.7 (*Loxsoma* R.Br., *Loxsomopsis* Christ)

FAMILY 20 Culcitaceae

Subfamily Culcitoideae Christenh.8 (*Culcita* C.Presl)

FAMILY 21 Plagiogyraceae

Subfamily Plagiogyrioideae Christenh.9 [*Plagiogyria* (Kunze) Mett.]

FAMILY 22 Cibotiaceae

Subfamily Cibotioideae Nayer (*Cibotium* Kaulf.)

FAMILY 23 Cyatheaceae Kaulf(approx14/approx 700)

Subfamily Cyatheoideae Endl.10 (*Alsophila* R.Br., *Cyathea* Sm., *Gymnosphaera Blume*, *Sphaeropteris* Bernh.)

FAMILY 24 Dicksoniaceae

Subfamily Dicksonioideae Link [*Calochlaena* (Maxon) R.A.White & M.D.Turner, Dicksonia L'He´r., *Lophosoria* C.Presl]

FAMILY 25 Metaxyaceae

Subfamily Metaxyoideae B.K.Nayar (*Metaxya* C.Presl)

ORDER N Polypodiales Link

FAMILY 26 Lonchitidaceae C.Presl (1/2) (*Lonchitis* L.)

FAMILY 27 Saccolomataceae Doweld11 (2/approx. 12) (*Orthiopteris* Copel., *Saccoloma* Kaulf.)

FAMILY 28 Cystodiaceae J.R.Croft (1/1) (*Cystodium* J.Sm.)

FAMILY 29 Lindsaeaceae C.Presl (6/approx. 220) [*Lindsaea* Dryander, *Nesolindsaea* Lehtonen & Christenh., *Odontosoria* Fe´e, *Osmolindsaea* (K.U.Kramer) Lehtonen & Christenh., *Sphenomeris* Maxon, *Tapeinidium* (C.Presl) C.Chr.]

FAMILY 30 Dennstaedtiaceae Lotsy (10/approx. 240) [*Blotiella* Tryon, *Dennstaedtia* Bernh., *Histiopteris* (J.Agardh) J.Sm., *Hypolepis* Bernh., *Leptolepia* Prantl, *Microlepia* C.Presl, *Monachosorum* Kunze, *Oenotrichia* Copel., *Paesia* St.-Hil., *Pteridium* Gled.]

FAMILY 31 Pteridaceae E.D.M.Kirchn., Schul-Bot.: 109 (1831).

Subfamily 31a Cryptogrammoideae S.Linds. Edinburgh J. Bot. 66(2): 358. 2009

(*Coniogramme*, *Cryptogramma*, Llavea).

Subfamily 31b Ceratopteridoideae (J.Sm.) R.M.Tryon, Amer. Fern J. 76: 184 (1986).

'Parkerioideae' of Smith et al. (2006a), nom. nud.

(*Acrostichum*, *Ceratopteris*).

Subfamily 31c Pteridoideae C.Chr. ex Crabbe, Jermy & Mickel, Fern Gaz. 11: 153 (1975).

Taenitidoideae (C.Presl) R.M.Tryon, Amer. Fern J. 76: 184 (1986).

(*Actiniopteris*, *Anogramma*, *Aspleniopsis*, *Austrogramme*, *Cerosora*, *Cosentinia*, *Jamesonia*, *Nephopteris*, *Onychium*, *Pityrogramma*, *Pteris*, *Pterozonium*, *Syngramma*, *Taenitis*)

Subfamily 31d Cheilanthoideae W.C. W.C.Shieh, J. Sci. Engin. (Nation. Chung-Hsing Univ.) 10: 211 (1973).

(Adiantopsis, Aleuritopteris, Argyrochosma, Aspidotis, Astrolepis, Bommeria, Calciphilopteris, Cassebeera, Cheilanthes, Cheiloplecton, Doryopteris, Hemionitis, Mildella, Notholaena, Paraceterach, Paragymnopteris, Pellaea, Pentagramma, Trachypteris, Tryonella).

Subfamily 31e Vittarioideae (C.Presl) Crabbe, Jermy & Mickel, Fern Gaz. 11: 154 (1975).

Adiantoideae (C.Presl) R.M.Tryon, Amer. Fern J. 76: 184 (1986).

(Adiantum, Ananthacorus, Anetium, Antrophyum, Haplopteris, Hecistopteris, Monogramma, Polytaenium, Radiovittaria, Rheopteris, Scoliosorus, Vittaria).

FAMILY 32 Cystopteridaceae Schmakov, Turczaninowia 4: 60 (2001).

4 genera (*Acystopteris*, *Cystoathyrium*, *Cystopteris*, *Gymnocarpium*).

FAMILY 33 Aspleniaceae Newman, Hist. Brit. Ferns: 6 (1840).

2 genera (*Asplenium*, *Hymenasplenium*).

FAMILY 34 Diplaziopsidaceae X.C.Zhang & Christenh., fam. nov.

3 genera (*Diplaziopsis* (type of the family), *Hemidictyum*, *Homalosorus*).

FAMILY 35 Thelypteridaceae Pic.Serm., Webbia 24: 709 (1970).

5 or more genera

FAMILY 36. Woodsiaceae Herter, Rev. Sudamer. Bot. 9: 14 (1949).

1–3 genera (*Cheilanthopsis*, *Hymenocystis*, *Woodsia*).

FAMILY 37 Rhachidosoraceae X.C.Zhang, fam. nov.

Based on a full and direct reference to the Latin description associated with Athyriaceae

Subfamily Rhachidosoroideae M.L.Wang & Y.T.Hsieh, Acta Phytotax. Sin. 42: 527 (2004).

1 genus (*Rhachidosorus* Ching, type of the family).

FAMILY 38 Onocleaceae Pic-Serm., Webbia 24:708 (1970)

1(-4) genera (*Onoclea*).

FAMILY 39 Blechnaceae Newman, Hist. Brit. Ferns, ed. 2:8 (1844).

Stenochlaenaceae Ching, Acta Phytotax. Sin. 16:18 (1978)

2-9 genera (*Blechnum, Brainea, Pteridoblechnum, Sadleria, Salpichlaena, Stenochlaena, Woodwardia*).

FAMILY 40 Athyriaceae Alston, Taxon 5: 25 (1956).

5 genera (*Anisocampium, Athyrium, Cornopteris, Deparia, Diplazium*)

FAMILY 41 Hypodematiaceae Ching, Acta Phytotax. Sin. 13: 96 (1975).

3 genera (*Didymochlaena, Hypodematium, Leucostegia*).

FAMILY 42 Dryopteridaceae Herter, Rev. Sudamer. Bot. 9: 15 (1949), nom. cons.

Aspidiaceae Mett. ex A.B.Frank in Leunis, Syn. Pflanzenk. ed. 2. 3: 1469 (1877), nom. illeg.

Filicaceae Juss., Gen. Pl.: 14. (1789), as 'Filices', nom. illeg., rej.

Peranemataceae Ching, Sunyatsenia 5: 208 (1940), nom. rej.

Elaphoglossaceae Pic.Serm., Webbia 23: 209 (1968).

Bolbitidaceae Ching, Acta Phytotax. Sin. 16: 15 (1978).

About 34 genera.

Subfamily 42a Dryopteridoideae B.K.Nayar, Taxon 19: 235 (1970).

(Acrophorus, Acrorumohra, Arachniodes, Ctenitis, Cyrtogonellum, Cyrtomidictyum, Cyrtomium, Diacalpe, Dryopolystichum, Dryopsis, Dryopteris, Leptorumohra, Lithostegia, Peranema, Phanerophlebia, Polystichopsis, Polystichum).

Subfamily 42b Elaphoglossoideae (Pic.Serm.) Crabbe, Jermy & Mickel, Fern Gaz. 11: 154 (1975). Based on Elaphoglossaceae Pic.Serm.

(Arthrobotrya, Bolbitis, Cyclodium, Elaphoglossum, Lastreopsis, Lomagramma, Maxonia, Megalastrum, Mickelia, Olfersia, Polybotrya, Rumohra, Stigmatopteris, Teratophyllum).

FAMILY 43 Lomariopsidaceae Alston, *Taxon 5: 25 (1956).*

3 genera (*Cyclopeltis, Lomariopsis, Thysanosoria*).

FAMILY 44 Nephrolepidaceae Pic.Serm., Webbia 29: 8 (1975).

1 genus (*Nephrolepis*).

FAMILY 45 Tectariaceae Panigrahi, J. Orissa Bot. Soc. 8: 41 (1986).

Dictyoxiphiaceae Ching, Sunyatsenia 5: 205, 218. 1940, nom. inval.

Hypoderriaceae Ching, Sunyatsenia 5: 209, 245. 1940, nom. inval.

6–10 genera *(Aenigmopteris, Arthropteris, Hypoderris, Pleocnemia, Psammiosorus, Psomiocarpa, Pteridrys, Tectaria, Triplophyllum, Wagneriopteris).*

FAMILY 46 Oleandraceae Ching ex Pic.Serm., Webbia 20: 745 (1965).

1 genus (*Oleandra*).

FAMILY 47 Davalliaceae M.R.Schomb., Reis. Br.-Guiana (Ri. Schomburgk) 2: 883 (1848).

2 genera (*Davallia, Davallodes*).

FAMILY 48 Polypodiaceae

Subfamily 48a Loxogrammoideae H.Schneid., subfam. nov. (*Dictymia, Loxogramme*).

Subfamily 48b Drynarioideae Crabbe, Jermy & Mickel, Fern Gaz. 11: 156 (1975).

(Aglaomorpha, Arthromeris, Christiopteris, Drynaria, Gymnogrammitis, Paraselliguea, Phymatopteris, Polypodiopteris, Selliguea).

Subfamily 48c Platycerioideae B.K.Nayar, Taxon 19: 233 (1970).

(*Platycerium, Pyrrosia*).

Subfamily 48d Microsoroideae B.K.Nayar, Taxon 19: 233 (1970), as 'Microsorioideae

Subfamily 48e Polypodioideae B.K.Nayar, Taxon 19: 234 (1970).

Another classification frequently used is by Smith et al., (2006) for extant ferns

The classification reflects recently published phylogenetic hypotheses based on both morphological and molecular data. In this system, 4 monophyletic classes, 11monophyletic orders, and 37 families, 32 of which are strongly supported as monophyletic besides a new family, Cibotiaceae Korall, are described. The phylogenetic affinities of a few genera in the order Polypodiales are unclear and their familial placements are therefore tentative. The classification also provides alphabetical lists of accepted genera (including common synonyms), families, orders, and taxa of higher rank are provided.

According to Smith et al., (2006) phylogenetic studies have revealed a basal dichotomy within vascular plants, separating the lycophytes (less than 1% of extant vascular plants) from the euphyllophytes (Fig. 2.1). Living euphyllophytes, in turn, comprise two major clades: the spermatophytes (seed plants), which are in excess with 260,000 species (Thorne, 2002; Scotland & Wortley, 2003), and the monilophytes (ferns, sensu Pryer et al., 2004b), with about 9,000 species, including horsetails, whisk ferns, and all eusporangiate and leptosporangiate ferns. Plants that are included in the lycophyte and fern clades are all spore-bearing or "seed-free", and because of this common feature their members have been lumped together historically under various terms, such as "pteridophytes" and "ferns and fern allies"—paraphyletic assemblages of plants.

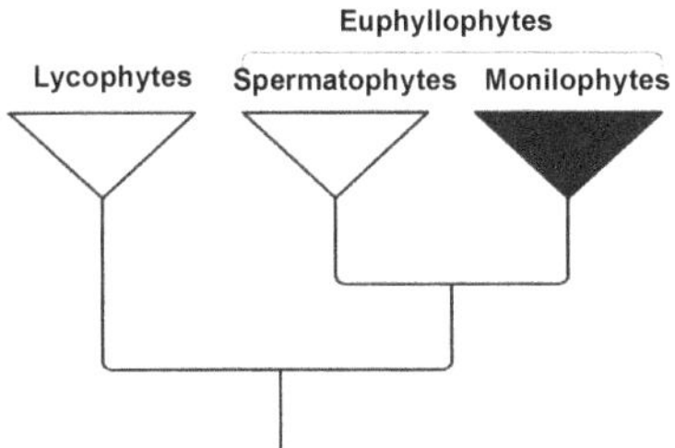

Figure 2.1: Phylogeny depicting relationships of major vascular plant lineages (after Smith et al., 2006).

The classification of Smith et al., (2006), proposed in the last eighty years may be taken as a comprehensive effort with data from previous major classifications taken into consideration.

I. CLASS PSILOTOPSIDA

A. ORDER OPHIOGLOSSALES

1. Family Ophioglossaceae. Ophioglossoids; incl. Botrychiaceae, Helminthostachyaceae. Four genera: *Botrychium* (grapeferns; moonworts), *Helminthostachys, Mankyua, Ophioglossum* (adder tongues). *Botrychium* (incl. *Botrychium* s.s., *Sceptridium, Botrypus,* and *Japanobotrychium*) and *Ophioglossum* (incl. *Cheiroglossa, Ophioderma*) are sometimes divided more finely (Kato, 1987; Hauk & al., 2003). Ca. 80 spp.; monophyletic.

B. ORDER PSILOTALES

2. Family Psilotaceae. Whisk ferns; incl. Tmesipteridaceae. Two genera (*Psilotum, Tmesipteris*), ca. 12 total spp. (2 in *Psilotum*); monophyletic (Hasebe & al., 1995; Pryer & al., 2001a, 2004).

II. CLASS EQUISETOPSIDA [= SPHENOPSIDA]

C. ORDER EQUISETALES

3. Family Equisetaceae. Horsetails. A single genus (*Equisetum*), 15 spp. usually placed in two well marked subgenera, subg. *Equisetum* and subg. *Hippochaete*; Monophyletic.

III. CLASS MARATTIOPSIDA

D. ORDER MARATTIALES — Including Christenseniales.

4. Family Marattiaceae. Marattioids; incl. Angiopteridaceae, Christenseniaceae, Danaeaceae, Kaulfussiaceae. Four genera: *Angiopteris, Christensenia, Danaea, Marattia*; *Archangiopteris* have been recognized by some.

IV. CLASS POLYPODIOPSIDA [= FILICOPSIDA]

E. ORDER OSMUNDALES

5. Family Osmundaceae. Three genera: *Leptopteris*, *Osmunda*, *Todea*. Ca. 20 spp.; monophyletic (Hasebe & al., 1995; Yatabe & al., 1999; Pryer & al., 2001a, 2004b).

F. ORDER HYMENOPHYLLALES

6. Family Hymenophyllaceae. Filmy ferns; incl. Trichomanaceae. Nine genera (Ebihara & al., 2006), two major clades (Pryer & al., 2001b), "trichomanoid" and "hymenophylloid", roughly corresponding to the classical genera *Trichomanes* s.l. and *Hymenophyllum* s.l. Ca. 600 spp.; monophyletic.

G. ORDER GLEICHENIALES

7. Family Gleicheniaceae. Gleichenioids, forking ferns; incl. Dicranopteridaceae, Stromatopteridaceae. Six genera (*Dicranopteris*, *Diplopterygium*, *Gleichenella*, *Gleichenia*, *Sticherus*, *Stromatopteris*), ca. 125 spp.; monophyletic.
8. Family Dipteridaceae including Cheiropleuriaceae. Two genera, *Cheiropleuria* and *Dipteris*, from India, southeast Asia, eastern and southern China, central and southern Japan, and Malesia, to Melanesia and western Polynesia (Samoa), ca. 11 spp.; monophyletic.
9. Family Matoniaceae. Matonioids. Two genera (*Matonia*, *Phanerosorus*), each with two spp.; monophyletic, sister to Dipteridaceae.

H. ORDER SCHIZAEALES — Monophyletic

10. Family Lygodiaceae. Climbing ferns. A single genus (*Lygodium*), ca. 25 spp.; monophyletic.
11. Family Anemiaceae. Including Mohriaceae. One genus (*Anemia*, incl. *Mohria*), ca. 100+ spp.; monophyletic.
12. Family Schizaeaceae — Two genera (*Actinostachys*, *Schizaea*), ca. 30 spp.; monophyletic.

I. ORDER SALVINIALES — Water ferns, heterosporous ferns; incl. "Hydropteridales", Marsileales, Pilulariales. Monophyletic common synonyms), families, orders, and taxa of higher rank are provided.

13. Family Marsileaceae — Clover ferns, incl. Pilulariaceae. Three genera *Marsilea*, *Pilularia*, *Regnellidium*), ca. 75 total spp.; monophyletic.
14. Family Salviniaceae — Floating ferns, mosquito ferns; incl. Azollaceae. Two genera (*Salvinia*, *Azolla*), ca. 16 spp.; monophyletic.

J. ORDER CYATHEALES — Tree ferns; incl. Dicksoniales, Hymenophyllopsidales, Loxomatales, Metaxyales, Plagiogyriales

15. Family Thyrsopteridaceae — One genus, *Thyrsopteris*, with a single species, *T. elegans*, endemic to the Juan Fernández Islands; clearly related to tree ferns, but of uncertain phylogenetic position within this group.
16. Family Loxomataceae — Two genera (*Loxoma*, *Loxsomopsis*), each with a single sp.; monophyletic.
17. Family Culcitaceae — One genus, *Culcita*, with two species; monophyletic.

18. Family Plagiogyriaceae — A single genus (*Plagiogyria*), with ca. 15 spp.
19. Family Cibotiaceae Korall, stat. nov. Based on a full and direct reference to the Latin description associated with subfam. Cibotioideae Nayar, Taxon 19: 234. 1970. – Type: *Cibotium* Kaulf.
20. Family Cyatheaceae — Cyatheoids, scaly tree ferns; incl. Alsophilaceae, Hymenophyllopsidaceae. Ca. five genera: *Alsophila* (incl. Nephelea), *Cyathea* (incl. *Cnemidaria, Hemitelia, Trichipteris*), *Gymnosphaera, Hymenophyllopsis, Sphaeropteris* (incl. Fourniera); 600+ spp.; monophyletic, together with Dicksoniaceae, Metaxyaceae, and Cibotiaceae constituting the "core tree ferns".
21. Family Dicksoniaceae, nom. cons. (Dicksonioids; incl. Lophosoriaceae). Three genera: *Calochlaena, Dicksonia, Lophosoria*). Ca. 30 spp.; monophyletic.
22. Family Metaxyaceae — A single genus (*Metaxya*), 2 spp.; monophyletic (Smith & al., 2001). Terrestrial, Neotropics.

K. ORDER POLYPODIALES — Including "Aspidiales"

23. Family Lindsaeaceae — Lindsaeoids; incl.Cystodiaceae, Lonchitidaceae. Ca. eight genera: *Cystodium, Lindsaea, Lonchitis, Odontosoria, Ormoloma, Sphenomeris, Tapeinidium, Xyropteris.*
24. Family Saccolomataceae — One genus, ca.12 spp.; apparently monophyletic, but more sampling is needed to determine whether the Old World species are congeneric with those from the New World. The relationships of *Saccoloma* (incl. Orthiopteris) have been contentious.
25. Family Dennstaedtiaceae — Dennstaedtioids; incl. Hypolepidaceae, Monachosoraceae, Pteridiaceae. Ca. 11 genera: *Blotiella, Coptodipteris, Dennstaedtia* (incl. Costaricia1), *Histiopteris, Hypolepis, Leptolepia, Microlepia, Monachosorum, Oenotrichia* s.s.1, *Paesia, Pteridium* (bracken). Ca. 170 spp.; monophyletic.
26. Family Pteridaceae — Pteroids or pteridoids; incl. Acrostichaceae, Actiniopteridaceae, Adiantaceae (adiantoids, maidenhairs), Anopteraceae, Antrophyaceae, Ceratopteridaceae, Cheilanthaceae (cheilanthoids), Cryptogrammaceae, Hemionitidaceae, Negripteridaceae, Parkeriaceae, Platyzomataceae, Sinopteridaceae, Taenitidaceae (taenitidoids), Vittariaceae (vittarioids, shoestring ferns). Ca. 50 genera, 950 spp. Constituent genera, some of them notoriously polyphyletic or paraphyletic and in need of redefinition. *Acrostichum, Actiniopteris, Adiantopsis, Adiantum, Aleuritopteris, Ananthacorus, Anetium, Anogramma, Antrophyum, Argyrochosma, Aspidotis, Astrolepis, Austrogramme, Bommeria, Cassebeera, Ceratopteris, Cerosora1, Cheilanthes, Cheiloplecton, Coniogramme, Cosentinia* (Nakazato & Gastony, 2001), *Cryptogramma, Doryopteris, Eriosorus, Haplopteris, Hecistopteris, Hemionitis, Holcochlaena, Jamesonia, Llavea, Mildella, Monogramma, Nephopteris1, Neurocallis, Notholaena, Ochropteris, Onychium, Paraceterach, Parahemionitis, Pellaea* (Kirkpatrick, unpubl.), *Pentagramma, Pityrogramma, Platyloma, Platyzoma, Polytaenium, Pteris* (incl. *Afropteris, Anopteris*), *Pterozonium, Radiovittaria, Rheopteris, Scoliosorus, Syngramma, Taenitis, Trachypteris*, and *Vittaria.*

27. Family Aspleniaceae — Asplenioids, spleenworts. 1-10 genera.
28. Family Thelypteridaceae. — Thelypteroids or thelypteridoids; incl. "Sphaerostephanaceae". Circa
29. Family Woodsiaceae — Athyrioids, lady ferns; incl. Athyriaceae, Cystopteridaceae. Circa ferns; incl. Athyriaceae, Cystopteridaceae. Circa 15 genera as defined here, ca. 700 spp., nearly 85% of them in the two main genera, *Athyrium* and *Diplazium* (incl. *Callipteris, Monomelangium*), which are both probably paraphyletic (Wang & al., 2003). Other widely recognized genera include *Acystopteris, Cheilanthopsis, Cornopteris, Cystopteris, Deparia* (incl. *Lunathyrium, Dryoathyrium, Athyriopsis*, and *Dictyodroma*; Sano.
30. Family Blechnaceae — Blechnoids; incl. Stenochlaenaceae). Currently ca. nine genera recognized (*Blechnum* s.l., *Brainea, Doodia, Pteridoblechnum, Sadleria, Salpichlaena, Steenisioblechnum, Stenochlaena, Woodwardia*).
31. Family Onocleaceae — Onocleoids. Four genera: *Matteuccia, Onoclea, Onocleopsis, Pentarhizidium*. 5 spp.; monophyletic, sister to Blechnaceae.
32. Family Dryopteridaceae — Dryopteroids or dryopteridoids; incl. Aspidiaceae", Bolbitidaceae, Elaphoglossaceae, Hypodematiaceae, Peranemataceae. Circa 40–45 genera, 1700 spp., of which 70% are in four genera (*Ctenitis, Dryopteris, Elaphoglossum*, and *Polystichum*).
33. Family Lomariopsidaceae — Lomariopsids; incl. Nephrolepidaceae, sword ferns. Four genera: *Cyclopeltis, Lomariopsis, Nephrolepis*, and *Thysanosoria*; ca. 70 species.
34. Family Tectariaceae — Tectarioids; incl. "Dictyoxiphiaceae", "Hypoderriaceae". 8–15 genera: *Aenigmopteris, Arthropteris, Heterogonium, Hypoderris, Pleocnemia, Psammiosorus, Psomiocarpa.*
35. Family Oleandraceae — Monogeneric, ca.40 spp., sister to Davalliaceae + Polypodiaceae.
36. Family Davalliaceae — Davallioids; excl. Gymnogrammitidaceae. 4–5 genera: *Araiostegia, Davallia* (incl. *Humata, Parasorus, Scyphularia*), *Davallodes, Pachypleuria*; ca. 65 spp. Monophyletic, sister to Polypodiaceae.
37. Family Polypodiaceae — Polygrams; incl. Drynariaceae, Grammitidaceae (grammitids), Gymnogrammitidaceae, Loxogrammaceae, Platyceriaceae.

    Pleurisoriopsidaceae. Ca. 56 genera, ca. 1200 spp. Pantropical, a few temperate Polypodiaceae s.s., as often recognized (e.g., by Kramer in Kubitzki, 1990), is paraphyletic.

Box 2.2

A recent scheme by Christenhusz and Chase (2014) which was also consistent with an understanding of fern phylogeny but represented a considerable departure in terms of stability, has not been widely adopted. However, Schuettpelz et al., (2016) have established the Pteridophyte Phylogeny Group (PPG), based loosely on the model employed for flowering plant classification (APG, 1998; APG II, 2003; APG III, 2009; APG IV, 2016). According to PPG, two pteridophyte classes: Lycopodiopsida (lycophytes) and Polypodiopsida (ferns) have been recognized. These are distinct lineages within the tracheophyte tree of life, with ferns resolved as more closely related to seed plants than to lycophytes (Kenrick & Crane, 1997; Pryer et al., 2001). Within Lycopodiopsida, further three orders (Lycopodiales, Isoëtales, and Selaginellales) have been recognized. Order Lycopodiales includes one family and 16 genera, whereas orders Isoëtales and Selaginellales each contain a single monogeneric family. Within Polypodiopsida, four subclasses have been identified:

Equisetidae (horsetails); Ophioglossidae; Marattiidae; and Polypodiidae (leptosporangiates). Extant Equisetidae includes a single order, a single small family, and a single genus (*Equisetum* L.). Subclass Ophioglossidae encompasses two orders, each with a single family, and a total of 12 genera. Marattiidae includes just one order, one family, and six genera. Subclass Polypodiidae comprises the vast majority of extant fern diversity. Seven orders (Osmundales, Hymenophyllales, Gleicheniales, Schizaeales, Salviniales, Cyatheales, and Polypodiales) have been identified, with the Polypodiales subsequently divided into six suborders (Saccolomatineae, Lindsaeineae, Pteridineae, Dennstaedtiineae, Aspleniineae, and Polypodiineae).Within Osmundales, a single small family with six genera is accepted. Order Hymenophyllales also includes just one family, but it is considerably larger and encompasses nine genera. Gleicheniales and Schizaeales each constitute three relatively small families, with ten and four total genera, respectively. The order Salviniales comprises two families and five genera, and the order Cyatheales encompasses seven small families, with a total of ten genera, plus the larger Cyatheaceae with three genera.

**Schuettpelz et al., (2016) have established the Pteridophyte Phylogeny Group (PPG), and have proposed a classification for all genera included time and again in Pteridophytes (Box 2.2). This classification is known as PPG 1 classification and is based loosely on the model employed for flowering plant classification (APG, 1998; APG II, 2003; APG III, 2009; APG IV, 2016). The author in the book has given common classifications for genera (included in Delhi University syllabus) whose detailed study is provided. It is however, recommended that since PPG 1 is inclusive of lycopods and ferns, it may be best followed for reference of the pteridophytes.

LYCOPODIUM

Lycopodiopsida (after Schuettpelz et al., 2016)

[LYCOPODS] (Christenhausz & Chase, 2014)

CLASS LYCOPODIOPSIDA Bartl. (**PPG 1, 2016**)

SUBCLASS LYCOPODIIDAE (Chase & Reveal, 2009; Christenhusz & Chase, 2014)

ORDER LYCOPODIALES (Christenhusz & Chase, 2014; **PPG 1, 2016**)

FAMILY LYCOPODIACEAE (Christenhusz & Chase, 2014; **PPG 1, 2016**)

Lycophytes or Lycopods belonging to a group of spore-bearing vascular plants, comprise close to 1,200 extant species. The lycophytes dominated the earth's landscape during the Carboniferous Period encompassing a tremendous expansion of terrestrial life roughly 360 million years ago (Stewart & Rothwell, 1993; Banks, 2009). Club mosses along with spike mosses (*Selaginella* sp.) are often referred to as 'living fossils', as they look very similar to their fossil relatives that lived 370-400 million years ago. Phylogenetically, the lycophytes are placed between the bryophytes and the euphyllophytes, which include the ferns, gymnosperms and flowering plants. Fossil records suggest that lycophytes, diverged from all other vascular plants including ferns and seed plants (euphyllophytes) more than 400 million years ago (Pryer et al., 2004; Fig. 3.1).

Lycophytes are key to understanding how major innovations evolved in order for plants to survive and thrive on land. These innovations include vascular tissue, leaves, stems, and lignification, traits that are important both in agriculture and as source of coal, and biofuel production. Selaginellaceae, together with the other two extant families Lycopodiaceae (club mosses) and Isoetaceae (quillworts) within the the class Lycopodiopsida, comprise the oldest lineage of vascular plants surviving on earth (Banks, 2009). The extant lycophytes, are typically small in stature, but many extinct lycophytes, such as the members of lepidodendrales (scaled trees), grew to enormous heights of about 40 meters (Stewart & Rothwell, 1993). Such giant lycophytes formed vast swamp forests, resulting in an interval of tremendous carbon fixation by terrestrial life, precipitating a significant drop of atmospheric CO_2 levels during the late Paleozoic era (Berner, 1993; Berner et al., 2000).

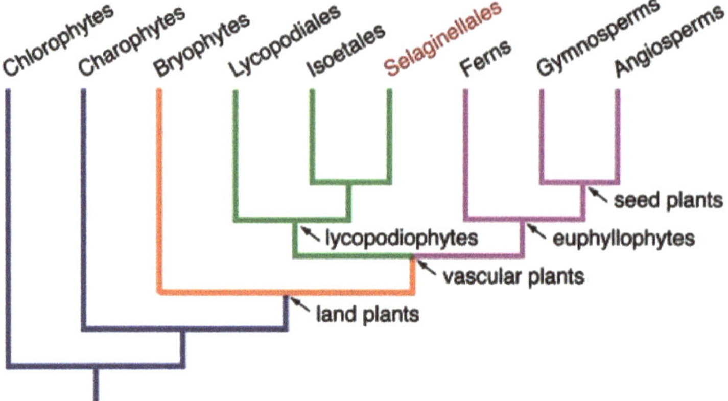

Figure 3.1: A simplified diagram illustrating the phylogeny of the green plant lineage. *Selaginella* is the only genus under the order Selaginellales, and represents an ancient lineage of vascular plants, Lycopodiopsida.

Lycopods differ from other vascular plants in having microphylls (simple leaves) with only one central vein and most of the members have subterranean gametophytes with an obligatory fungal association. Lycopods with these subterranean gametophytes need some disturbance of the ground cover to disperse the spores into the soil, which is why they are successful in colonizing abandoned gravel and sandpits or soil following forest fires.

Classification based on the Pteridophyte Phylogeny Group I (PPG I) system, published in 2016 subdivides Lycopods into three families: Lycopodiaceae (lycopodiales, clubmosses), Selaginellaceae (Selaginellales, spikemosses) and Isoëtaceae (Isoetales, quillworts).Terrestrial Selaginellaceae and aquatic Isoëtaceae are unique in having heterospory (large and small spores), where the gametophyte develops inside the spore wall. This is possibly an adaptation to protect small gametophytes from adverse environmental conditions. Endosporic germination can also be found in families of aquatic ferns - Marsileaceae and Salviniaceae. Lycopods are widely distributed in world but are especially numerous in the tropics.

LYCOPODIACEAE (CLUB MOSSES)

Commonly known as Club mosses, the plants are low evergreen herbs with needle-like or scale-like leaves. Many species possess cone-like clusters of small leaves (sporophylls), each with a kidney-shaped spore capsule at its base and form strobili. The plants are homosporous, meaning they produce just one kind of spore. The terrestrial or subterranean gametophytes vary in size and shape depending on the genera. In some species, including nearly all those of the North Temperate Zone, the subterranean gametophyte is associated with a fungus for continued growth. This sexual phase (gametophyte) alternates in the life cycle with the spore-producing plant (sporophyte), which is above ground and is formed upon fertilization.

In 2016, the Pteridophyte Phylogeny Group 1 recognized 16 genera in the family Lycopodiaceae including *Huperzia* (10–15 species), *Lycopodiella* (15 species), and *Lycopodium* (9–15 species) based on the characters of branching patterns, the presence or absence of modified leaves, spore morphology, gametophyte shape and chromosome number.

Huperzia lucidula (Shining club moss), a North American species inhabits wet woods and rocks, has no distinct strobili; it bears its sporangia at the bases of leaves scattered along the branches. Fir club moss (*H. selago*), a tall plant about 20 cm (8 inch) in height and is native to rocks and bog margins in the Northern Hemisphere. It also lacks distinct strobili. Common club moss -*Lycopodium clavatum*, also known as running pine or stag's horn moss, has creeping stems to 3 metres (about 10 feet) long and 10 cm (about 4 inch) high ascending branches. The scale-like green leaves are closely set together. The plant inhabits open, dry woods and rocky places in the Northern Hemisphere. The sporophylls (spore-producing leaves) are arranged in pairs along a stalk-like strobilus. *Dendrolycopodium obscurum* (ground pine), a 25 cm (10 inch) tall plant, has underground-running stems. Native to northern North America, to mountain areas farther south, and to eastern Asia, it grows in moist woods and bog margins. *Diphasiastrum alpinum* (Alpine club moss), with yellowish or grayish leaves, is native to cold woods and Alpine mountains in northern North America and Eurasia while *D. digitatum* (ground cedar) producing fan-like branches resembling juniper branchlets is native to northern North America. Members of the genus *Lycopodiella* commonly known as bog club moss, inhabit bogs and other wetland habitats nearly worldwide, and many species are native to tropical areas of the Americas. *Phylloglossum drummondii* (pygmy club moss) is the only member of its genus and is found in parts of Australia and New Zealand.

LYCOPODIUM

It is a large genus, growing in arctic, temperate and tropical lands but chiefly in subtropical and tropical forests. It is not found in arid regions.

The club mosses of northern temperate climates are chiefly low, evergreen, trailing plants of woodlands and barren fields. The plants are characterized by the small moss-like leaves and the club- shaped fruiting stem tips or strobili. The bushy, upright branches of some species look similar to minute pine trees and the plants are therefore sometimes also known as 'ground pines'. The general appearance and the leaves, and strobili seen are fairly typical of the group.

SPOROPHYTE

Morphology

- The species generally vary greatly but all are slender and weak stemmed.
- Many species are prostrate where the main stems creep on or below the surface of the ground; some are semi-erect or ascending where the older

parts of the stem become finally horizontal.

- The stems of some species scramble over shrubs, twining to some extent while many of the tropical species are epiphytes with weak pendent branches (**PLATE 1A,B**).

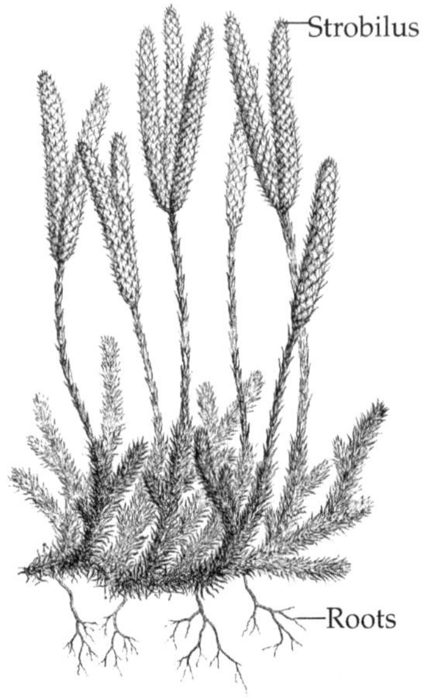

Figure 3.2: *Lycopodium* (Morphology)

The plant has creeping stem with erect tips. Stems are dichotomously branched, one of the branches remains small and the other grows to a greater length. Smaller and erect branches bear cone or strobili at the tips. Basal surface of creeping stems bear dichotomously branched and thin adventitious roots that abound all along its length.

Stem

- The stems clothed with abundant small leaves, branch freely characteristically dichotomously (Fig. 3.2).
- The successive forkings are usually in planes at right angles with one another, but may be obscure occasionally.
- The two branches may be equal and alike, and may continue to grow indefinitely as in typical dichotomy or one branch of the dichotomy may be weaker than the other, thus simulating a monopodial condition but not truly monopodial.
- Except in cases where the branching is by symmetrical dichotomies, most of branches produced out of unequal branching, become lateral structures of limited growth upon a rhizome-like main stem. These determinate lateral branches which branch freely dichotomously in one plane or sparsely, have

a limited period of growth (one or few seasons). The limited and often elaborate branch systems are photosynthetic and produce fruiting structures (see Box 3.1).

Leaves

- Small leaves – microphyllous, typically 2 to 10 mm long, though a few species have somewhat larger leaves, at most 25 to 35 mm long simple, numerous, closely placed covering the stem especially during strobilus formation (Fig. 3.2).
- The arrangement is in close spirals, whorls or opposite pairs or is somewhat irregular, in some species it is said to be without order.
- Where the leaf arrangement is definite, the phyllotactic fractions are unlike those of other vascular plants, such series as 2/7, 2/9, 2/11.
- In the creeping species, leaves of the determinate lateral branches are much specialized. The phyllotaxy is decussate, a leaf arrangement commonly recognized as one of the highest type. The leaves are fused to the branchlets and decurrent (curved downwards) and are photosynthetic structures suggesting frond-like leaves. The leaves of the lateral rows are enlarged and wing-like, those of others are reduced, while the lower dorsal row is abortive. These dorsiventral branchlet systems resemble the twigs of Thuja, a gymnosperm.

Root

- The first root that develops as the young sporophyte becomes independent of the gametophyte, is small and generally short lived.
- It is replaced in older plants by adventitious roots that arise singly or in groups acropetally along the underside of the stem (Fig. 3.2). Such roots are somewhat irregularly distributed and in position and bear little or no relation to other organs.
- The roots are strikingly dichotomous in some species, for example, in *L. obscurum* and *L. lucidulum,* with each successive forking at right angles to the preceding one while in other species the dichotomy is obscure.
- The roots do not become extensive and no endogenous lateral rootlets arise as in seed plants. However, root hairs in terrestrial forms are abundant and persist over a long period.

Anatomy

Stem

- The stem has one celled epidermis with stomata, a well differentiated cortex and a central vascular cylinder (Fig. 3.3A).
- Relative thickness of cortex is variable in different species. While in some species with small stem, cortex remains parenchymatous, in others, the cortex at maturity gets sclerified into outer and inner zones, forming

three bands - hypodermal chlorenchyma, followed by parenchyma, and sclerenchyma that encircle stele, and in very old stems, the entire cortex gets sclerified. In subterranean stems, the cortex is often completely sclerenchymatous while the aerial stems often have large air spaces in cortex due to disruption of cells.

Box 3.1

Usually the descriptions of the vegetative body of *Lycopodium* apply a number of terms which were originally coined for plants with lateral branching, such as sympodial and monopodial branching, main and lateral branches, or principal and secondary branches. But with present classification, these terms have become redundant. However, terms Urostachya and Rhaphalostachya which have existed in literature for a long period, have been explained here.

Subgenus Urostachya possesses branched or pendant but never creeping members. This subgenus includes *L. lucidulum, L. phlegmaria* and others where branching is always dichotomous, usually at right angles to one another. The species bear the adventitious roots along the stem. All leaves are similar and sporophylls are similar to vegetative leaves and not aggregated into strobili. Adventitious shoots may be formed by lateral branching at the base of senile or damaged plants in some terrestrial or epiphytic species, and so may rejuvenate them.

Subgenus Rhaphalostachya includes *L. inundulatum, L. complanatum, L. cernuum.* The members possess prostrate branching of the stem in the first-formed portion while later-developed portion it is always monopodial. In these species, the first time considerable modifications in the structure of the sporophyll are seen. Though the external difference between the sporophyll and the vegetative leaf is that sporophyll is of considerable thickness and has a prominent dorsal ridge. This ridge projects from the back of the sporophyll and overlaps the sporangium belonging to the sporophyll immediately below. The sporangium is thus far less exposed than it is in the Urostachya, but it is still visible on the exterior of the strobilus.

- Endodermis is usually ill-defined and can be identified in younger portions due to thickened radial-walls. Internal to endodermis is pericycle layer of three to six cells.
- Stelar anatomy differs from species to species and even within the species. However, young plants in all species conform to a single pattern – protostele, exhibiting a single mass of exarch xylem with radiating arms that enclose phloem. The xylem and phloem remain separated from each other by a layer of stelar parenchyma.
- As the relations of xylem and phloem are unusual between and within species, complex steles result. In the simpler types, the xylem forms a star-like mass with rays of varying number and is said to be actinostele. In *L. phlegmaria* exarch actinostele is simple but in *L. serratum*, xylem rays instead of being pointed are expanded outwards (Fig. 3.3B). In the bays or

arms of the star lies phloem, separated from the xylem by a narrow layer of parenchyma. In more complex types, the furrows in the xylem cylinder are numerous and may break up the xylem into isolated strands (Fig. 3.3C). These apparently separate xylem strands are however, free only at certain levels and do not represent discrete vascular bundles. The furrowing may be such as to break up the xylem core into plate-like lobes as in *L. volubile* which is plectostele (Fig. 3.3D) or into a mesh-like mass or as scattered groups with included bands of phloem – mixed protostele as in *L. cernuum* (Fig. 3.3E). From the ridges, leaf traces extend as slender strands obliquely upward and outward through the pericycle and cortex to the leaves. One strand enters the base of each leaf to become an unbranched mid vein. The xylem is simple, with scalariform pitting in metaxylem. The phloem is also simple consisting of sieve tubes and parenchyma cells **(PLATE 1C-F)**.

- The stem is wholly primary in structure, no cambium arises to develop secondary xylem and phloem and no periderm is formed.

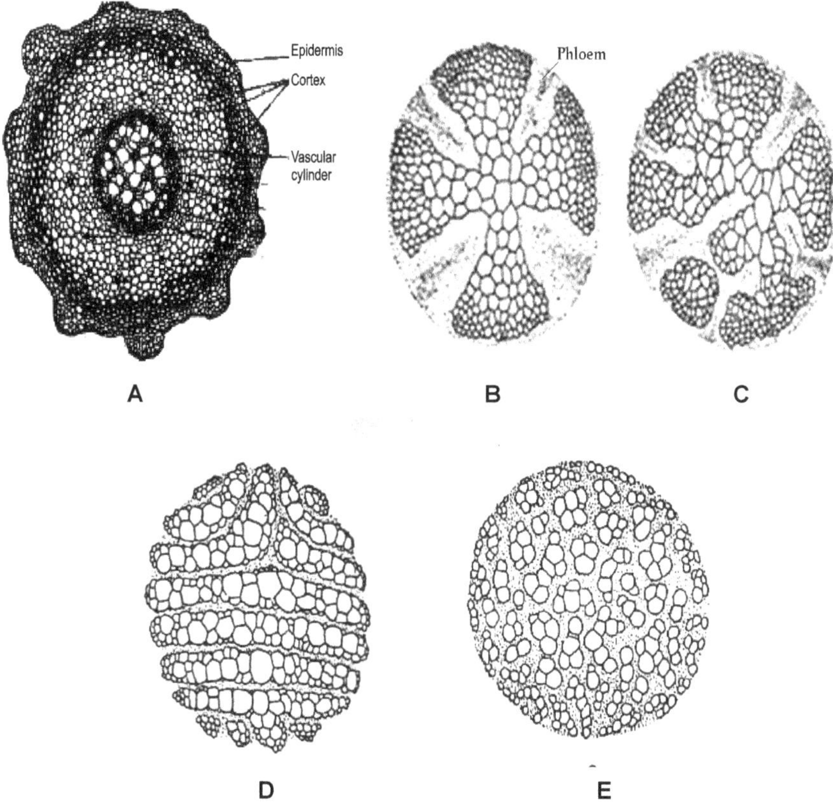

Figure 3.3A-E: *Lycopodium* sp (Anatomy)

A -T.S Stem – *L. cernuum*. The epidermis is followed by a large cortex which is differentiated into three distinct zones – outer chlorenchymatous, middle is sclerified and innermost is parenchymatous. Endodermis and pericycle are ill-defined while the mixed protostele in which mesh-like mass of xylem remains surrounded by phloem, represents the vascular strand. B-E *Lycopodium* sp – B- The

stelar region of various species enlarged to show actinostele with fan shaped free ends of xylem. In the arms lies phloem (*L.serratum*). In *L. lucidulum* the furrows in xylem cylinder are more numerous, breaking up xylem into isolated strands (C). They do not form discrete bundles as they remain attached at several points. In *L. cernuum*, furrowing may lead to formation of plate-like lobes as in *L.volubile* (D) or mesh-like xylem masses (E).

Root

- Epidermis, the outermost layer shows root hairs and is followed by a large cortex which is differentiated usually into outer thick walled region meant for mechanical strength (Fig. 3.4A,B).
- Stele in young sporophyte is represented by a monarch xylem with protoxylem as one mass which is exarch to decarch in mature roots. In *L. selago* xylem is 'c'-shaped and two arms of 'c' have protoxylem while metaxylem is centrally placed, phloem is placed between two protoxylem masses. The protoxylem and metaxylem elements are similar to those of the stem (Fig. 3.4C).

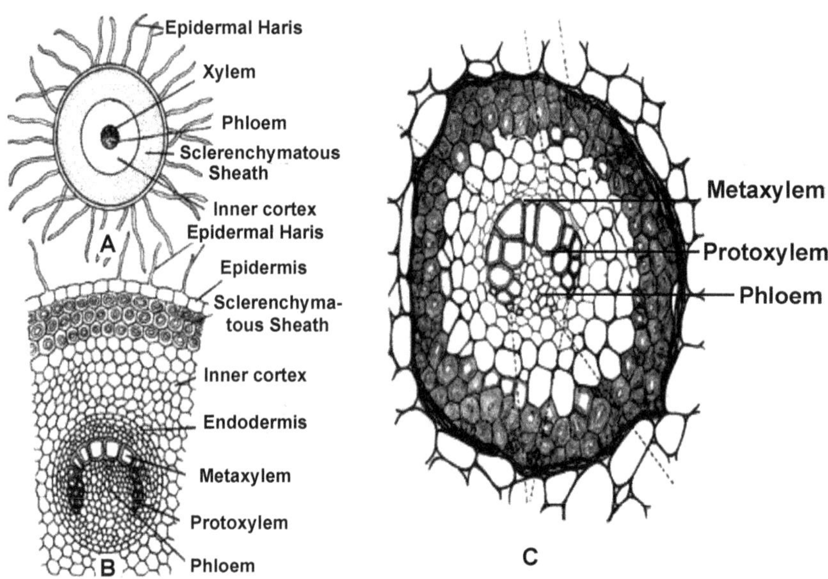

Figure 3.4A-C: *Lycopodium* (Anatomy)

T.S. Root – A – is an outline diagram. The epidermis shows root hairs and is followed by a large cortex which has a few outer layers thickened for mechanical support (B). The protostele is monarch and 'C' shaped – with protoxylem towards two arms of 'C' and metaxylem in the centre (C).

Leaf

- As the leaves are small, the structure is simple with cuticularized epidermis possessing stomata followed by a well-defined mesophyll.

- A single median vascular bundle, concentric in structure with no distinct protoxylem and metaxylem is present. The xylem with annular and spiral thickenings is surrounded by a zone of phloem with scattered sieve tubes.

Reproduction

Vegetative reproduction occurs by several means (Fig. 3.5A-C).

- Bulbils and gemmae formed on the new stem tips in several species when fall on ground, they root at once. They are thought to be flattened tips of lateral branchlets with enlarged wing-like leaves.
- In other cases, the rhizome may continue to grow indefinitely at the apex while the older part dies. In this way as branching regions are passed in the dying off, the original plant is broken up into individuals.
- In *L. inundatum* during winters, or at the beginning of the season's growth, the entire plant dies except the tip of the rhizome which becomes a resting bud. However, this is essentially an annual habit.
- In epiphytic species, fragments of the plant body have the potential to produce new plants freely. Such species also produce two types of reproductive bodies, minute ones which rest before development and larger ones which grow at once.
- Vegetative propagation of gametophytes is suggestive of liverworts and mosses rather than vascular plants.

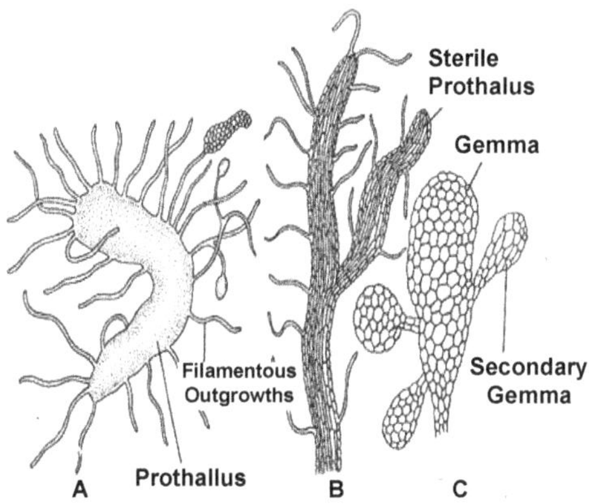

Figure 3.5A-C: *Lycopodium* (Vegetative reproduction)

The gametophytes show gemma, the vegetative propagules.

Reproduction by spores

- The sporophylls are aggregated into definite cone-like structures or strobili (Fig. 3.6A, **PLATE 1G**).
- One of the characteristics of Lycopsida is the association between sporangium and sporophyll; each sporangium is generally either located on the adaxial

side of the sporophyll or in its axil. In certain cases sporangia may be (cauline) on the stem just above the leaf (Fig. 3.6B-E).

- Species like *L. selago* and *L. lucidulum*, have sporophylls similar to vegetative leaves and no definite strobili are organized. There are however, fertile areas on the stem that alternate with the vegetative or sterile regions.
- In advanced species, the sporophylls are aggregated into definite cone-like structures or strobili and the sporophylls of such strobili may be quite unlike vegetative leaves in size, shape and colour. They may also exhibit other specializations related to sporangial protection and spore dispersal. These strobili may occur on leafy stems or they may be elevated on lateral branches with stalk-like structures which possess very small scale-like leaves unlike those of the vegetative shoot.

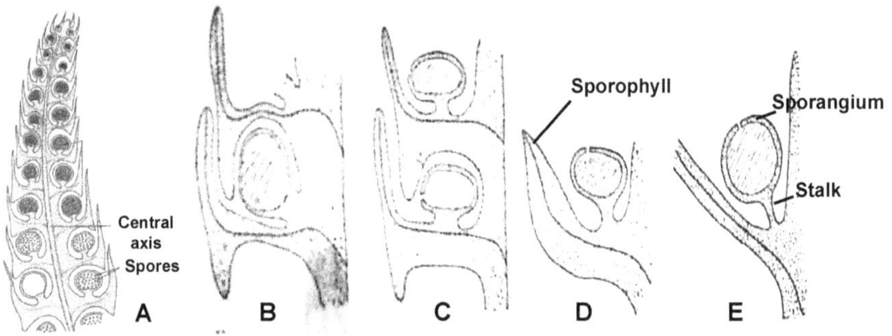

Figure 3.6A-E: *Lycopodium* (Strobilus and Sporangial position)

L.S of reproductive stem to show central axis, sporophylls and sporangia (A). B-E - position and point of sporangial dehiscence. Sporangia are axillary and protected as in *L. inundatum*, (B). (C) – foliar and protected as in *L. cernuum*, (D) – sub foliar and exposed as in *L. squarrosum* or axillary and exposed as in *L. lucidulum* (E).

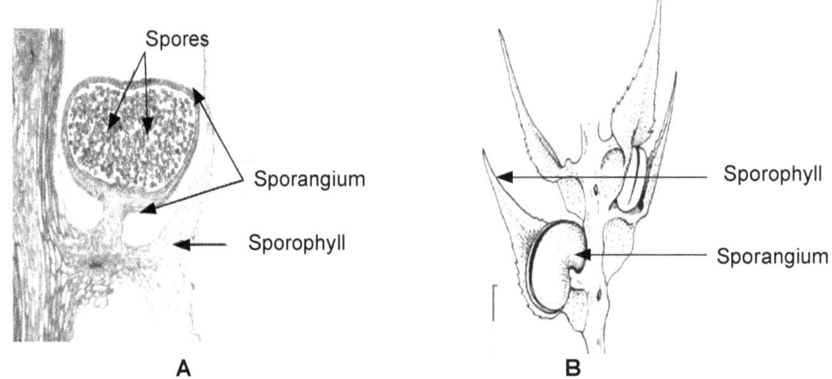

Figure 3.7A, B: *Lycopodium* (Sporangia)

A - Sporangia are borne on the sporophylls, are stalked, have a subpeltate exterior face and remain protected by sporophyll above (B).

Sporangia

- In the strobili bearing species, sporangia may be largely or wholly enclosed by dorsal lobes of the sporophylls above (Fig. 3.7A).
- Development of the sporangium is of the eusporangiate type. A group of surface cells divide periclinally forming outer cells that divide periclinally and anticlinally resulting in multi-layered wall, and inner derivatives that result in the sporogenous cells.
- The innermost layer of the sporangial wall functions as the tapetal layer.
- The sporangia are large (1.0 to 2.5 mm in diameter), reniform or in some cases subspherical, yellowish when mature, and possess a short stalk or pad-like base (Fig. 3.7B, **PLATE 1H**).
- When mature they open by a slit transverse to the leaf into two clam-like valves and the spores (homosporous) are scattered by the wind. The spores bear a wall which may or may not be smooth.

Gametophytes

- The genus *Lycopodium* has spores that do not germinate for a very long time and gametophytes or prothalli developed subsequently require about two years to mature. As a result, the prothalli cannot be cultivated readily. However, prothalli of both tropical epiphytic and the temperate terrestrial species are of fairly common occurrence.
- There are two distinct types of gametophytes. The first type is green except at the base from where rhizoids arise and develops on the surface of the ground. It is very small only 2 or 3 mm long, cylindrical or ovoid with a lobed branching top. Sex organs generally occur near the base of aerial lobes (Fig. 3.8A,B). The spores of these species germinate in a few days and the gametophyte matures quickly usually in one season. But the gametophytes are short lived.
- The other type is much larger, 1 or 2 cm long or wide, is non green and subterranean, and tuberous. It may appear in some species like a top, or a carrot (Fig. 3.8C), while in others, it resembles a disk with convoluted margins. Germination is long delayed, in some species it may be as late as three to five years while in others it may take six to eight years. It is believed that this slow growth is because until the gametophyte gets infected with the fungus, it cannot grow (Fig. 3.8D). The dependence on fungus is perhaps physiological and it may provide the gametophyte with some minerals that are required for growth and if the prothallus is not infected, its growth ceases. The gametophyte of such species grows slowly, taking several (6 to 15) years to mature and is long lived. The sex organs are segregated into definite groups but in course of development, antheridia appear first near the middle of the crown of prothallus. Such gametophytes continue to nourish the young sporophyte even after maturity.
- The first type is found in tropical species for example, *L. cernuum*, less commonly in temperate forms as in *L. inundatum*.

- The second type is characteristic of northern creeping species such as *L. obsurum, L. complanatum* and *L. clavatum* and is also seen in epiphytic forms where the gametophytes are found under humus on tree trunks and where they remain attached to the substratum by absorbent hairs. In such species the prothallus may be cylindrical, branched, tuberous, with branches extending in various directions. The whole thallus is heavily infested with fungus and sex organs are borne on extended upper surface (**PLATE 1I,J**). The outermost layer is cutinized epidermis and bears rhizoids which are extensions of epidermal cells. The epidermal cells are somewhat flattened and do not show presence of fungus. Inner to single layered epidermis is a band of three to four layers of cells, which are extended parallel to the surface of the prothallus. These cells, as well as the succeeding single layer which consists of more or less regular palisade-shaped, are filled with the mycelium of the intracellular fungus. They appear darker than the other regions which lack fungus. Above the layer of palisade cells comes a region about eight cells deep, from the cell-cavities of which the fungus is absent. These cells are smaller than those of the preceding layer, but are not infrequently elongated in the same direction. Their walls appear thicker than those of the rest of the prothallus, especially at the angles. Throughout the whole of this region the mycelium of the fungus can be traced within these thickened walls, but never penetrating into the cavities of the cells. The latter contain a large quantity of starch, which is also present, though less abundantly, in the tissues above, but is practically absent from those in which the mycelium is intracellular. The cells composing it are thin-walled and possess a scanty protoplasmic lining. Those situated more internally are of large size, but cells surrounding the antheridia or archegonia are much smaller. This uppermost layer may be distinguished as the generative layer, while the large-celled tissue beneath maybe compared to the cushion of a fern-prothallus. The prothallus is, thus seen to consist structurally of six more or less clearly defined regions, the four lowest of which are, as their structure indicates, concerned with the nutrition of the organism, while the uppermost layer contains the sexual organs.
- There also occur intermediate forms between the two types, and such gametophytes are lesser known. For example in *L. selago*, the prothallus is usually short, upright, and cylindrical with a meristematic zone and is devoid of chlorophyll (Fig. 3.8B). The sex organs are borne on the upper surface. But if this prothallus gets closer to the ground, the apex emerges and the prothallus starts acquiring chlorophyll and from radially symmetrical, it becomes dorsiventral. Certain regions of the margin grow out into lobes that bear sex organs on the upper surface while lower bears rhizoids.
- The prothallus of *L. selago* has been described by Bruchmann to present some exceedingly interesting modifications of form. According to him "This variety of form of the prothalli seems to be dependent mostly on the soil in which they are produced. The elongated cylindrical forms are found especially in firm soil, in which they strove towards its surface mostly in a vertical direction. In loose soil, especially near the surface, flat forms of prothalli are

seen." Besides the subterranean forms of prothallus of this species, he found some which grew wholly or partly at the surface of the earth, and which in their upper part showed a thoroughly green colouring. Such prothalli lived as semi-saprophytes, and represent "an interesting transition between the assimilating and the merely saprophytic forms" of the *Lycopodium* prothallus. Every complete prothallus, whether of the elongated or thickset form, shows at the original end a tiny, usually bent point, which is the region immediately developed from the spore, and which develops above into the cone-shaped tissue-body. The prothallus is abundantly supplied with long rhizoids; and many-celled paraphyses, similar to those described by Treub in *L. phlegmaria*, are present, along with the sexual organs. In discussing the different types of *Lycopodium* prothallus, Lang with regard to prothallus of *L. selago* states that "The two forms of prothallus found in *L. selago* give the clue to the more specialized saprophytic types, which in the deeper-growing subterranean species retain the radial symmetry while becoming modified in shape. On the other hand, the type of prothallus growing in rotting wood has lost the radial symmetry, and consists of cylindrical but more or less clearly dorsiventral branches." He gave to the prothallus of this species the rank of a new type, and concluded a separation of the lycopodiums into groups, or, better still, into genera. However, the facts known concerning the gametophyte generation of the species which comprise the subgenus *Urostachya* undoubtedly point to the fact that the two sections *selago* and *phlegmaria* are closely related. *L. selago* is probably to be regarded as the primitive type of the subgenus. The *phlegmaria* type of prothallus, found to occur in all the main divisions of the subgenus where the species have an epiphytic habit, has arisen as a modification of the *selago* type in accordance with that habit of growth. It is further concluded that evolution in the form of the prothallus in the subgenus *Urostachya* has proceeded along several parallel lines, and that there are to be traced, as Lang suggests "instances of independent adaptation to similar conditions."

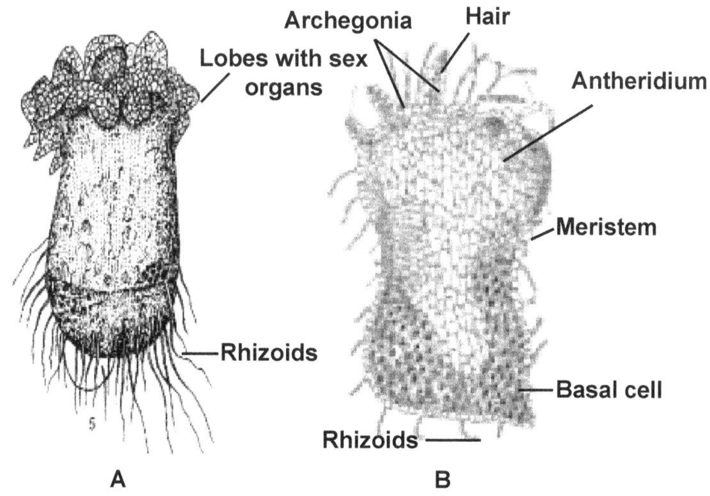

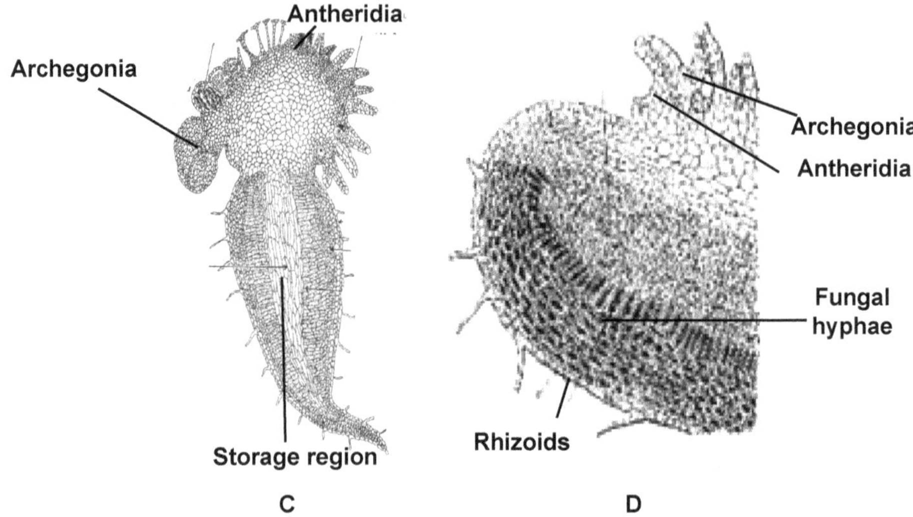

Figure 3.8A-D: *Lycopodium* (Range in Prothallus)

A. In *L. cernuum*, the prothalli are very small with some portion buried in ground (A). In *L. selago* also, commonly the gametophytes are with same general plan, though they may be smaller in size. The prothalli are marked with a meristematic region just below the free end and bears reproductive structures for sexual reproduction (B). In *L. clavatum*, the prothalli are carrot shaped. The endophytic fungus is present in the cortical region, the upper lobed portion bears sex organs – archegonia on the margins and antheridia in the centre (C). Internally this prothallus has a central region of elongated cells while the cortical region has endophytic fungus (D).

Sex Organs

- The gametophytes are monoecious with antheridia developing first.
- The numerous antheridia and archegonia are borne on the crown/top or at the base of its lobes or arms when these are present, which depends on the species being observed.
- On the flattened type of prothallus, sex organs are borne on the central cushion which apparently represents the crown of the elongated radial type. In this type, the first sex organs are antheridia only, which are borne in the centre of or all over the cushion. Later both antheridia and archegonia are borne around the edges of the cushion which enlarges as the gametophyte increases in diameter.
- Here the two kinds of organs are not intermingled but are borne close together in large clusters; in the green type prothallus they may be intermingled. On subterranean prothalli mature sex organs are present at all times of the year.

- The antheridia are somewhat indefinite structures, varying in size, shape and number of antherozoids even in the same plant. They are either wholly sunken in the gametophyte or project slightly. The opercular cells are prominent. In each antheridium a large number of biflagellate antherozoids are formed (more similar to those of bryophytes than vascular).
- The archegonia are sunken with only necks protruding. In subterranean type of gametophyte archegonia are slender, cylindrical structures with a long neck of about 6-14 cells, while those of the surface-living type are short with three or four tiers of neck cells. The venter is narrow with an egg cell, and a venter canal cell.

Young Sporophyte

- The embryo is endoscopic with shoot directed away from the archegonial neck.
- The zygote undergoes a division transverse to the long axis of the archegonium. Of the two cells formed, the outer forms a suspensor which plays the role of restricting the embryo proper to the inner half of the archegonium rather than of anchorage and absorption.
- The suspensor later forms a subspherical foot, an organ of absorption of food from the prothallus.
- Soon stem and first leaf of the sporophyte is organized from the other cell or the sister derivative of suspensor, followed by organization of the root which does not become extensive, is variable and exogenous (Fig. 3.9A,B).
- The first erect stem and first root are both short lived in some species, and new adventitious roots arise.
- In species with green prothallii as in *L. cernuum,* an important variation occurs in embryo development. The foot formed from the cells near the suspensor remains small and the other segment derivatives which in most species form leaf and shoot apex, do not organize into organs in this species. Prothallus continues to divide in various planes forming a spherical parenchymatous neutral body, termed protocorm which pushes its way through the gametophyte. This body is provided with rhizoids and has on its upper part, a few cylindrical, green leaf-like projections or protophylls indefinite in number and arrangement. There is an associated fungus as in the gametophyte. Later stem, leaves and adventitious roots emerge from the protocorm but there is absence of vascular tissue in protocorm. The stem-axis arises at some point on the dorsal surface and at the same time, vascular tissues are initiated from the stem apex. With the differentiation of the stem-axis both the protophylls and the rhizoids decay away.
- Soon the sporophyte grows into a well differentiated plant.

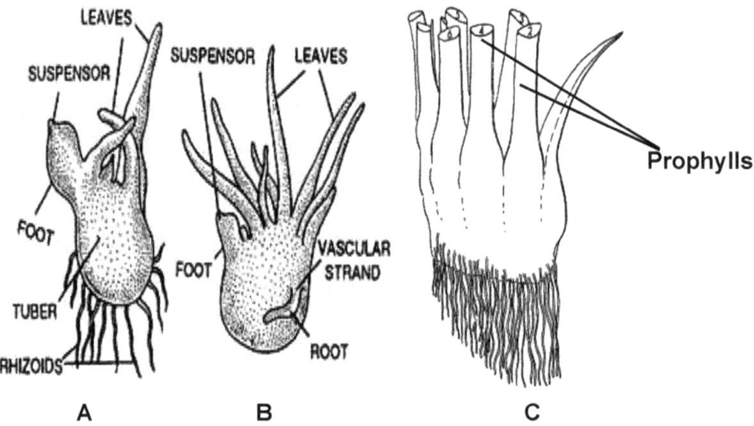

Figure 3.9A-C: ***Lycopodium*** **(Embryo)**

The development is endoscopic and shows a suspensor and foot in initial stages (A) while the stem and leaves get organized a little later (B). C-Protocorm with several prophylls.

**The protocorm (Fig. 3.9C) has been interpreted as a stage of sporophyte development interpolated between gametophyte and a typical adult sporophyte. Seen in Lycopodium cernuum, and also in L. inundatum* and *L. salakense, the development of the tuberous organ termed the "protocorm" is an important phase in the ontogeny, bridging over the period between the early stage in which the embryo plant is wholly dependent upon its parent prothallus, and its subsequent development, in which it has obtained independence through the establishment of a root-system.*

Evolutionary interpretations

- The characteristic dichotomous branching of the stem in some species of *Lycopodium* is suggestive of lower plants rather than of vascular plants.
- The various arrangements of leaves show a series of progressive specialization – from indefinitely placed arrangement to spiral to whorled and finally opposite, the last being more specialized and found in other groups of plants.
- The first leaves are scale-like and without vein, thus leaves in this group are believed to have arisen as enations first and vasculature was acquired later.
- The dichotomy seen in stems continues in the roots where all roots are adventitious. The first formed roots are short-lived and there are no endogenous roots. They are therefore, not a fundamental part of sporophytic body but are secondary structures. The embryo lacks a root quadrant.
- The stem vasculature is generally protostelic with exarch xylem which is a primitive condition. However, a great variation is seen within and among species.
- The branch system in dorsiventral type shows lower abortive row, and many a times shows lateral branches of limited growth, a character of gymnosperms with branches of determinate growth.

- The sporophylls in early genera are not organized in strobili and that they resemble vegetative leaves. Later they organize into strobili and morphological differences between vegetative and sporophylls are also of evolutionary advancement where the later type provides protection to sporangia.
- Presence of foliar and cauline sporangia and the intermediate forms and presence of green gametophytes to subterranean non green types where both have fungal association, are also characteristic of the group.
- The elongate archegonium with several neck canal cells stands as most primitive among vascular plants.
- The suspensor though not fully developed yet plays the true role of restricting embryonic growth.
- The primitive sporophyte in some species remains dependent on gametophyte as all roots are adventitious and in some species even these develop late.
- The protocorm is undoubtedly a structure that helps delicate sporophyte tide over unfavourable conditions.
- Closely similar forms are found among fossils of Carboniferous period indicating the primitiveness of the genus.

Class	LYCOPODIOPSIDA
Order	LYCOPODIALES
Family	LYCOPODIACEAE
Sub Family	LYCOPODIOIDEAE
	LYCOPODIUM

QUESTIONS

Q1. Bring out the differences between:

(a) Urostachya and Rhaphalostachya

(b) Plectostele and siphonostele

Q2. Write short notes on:

(a) Protocorm

(b) Evolutionary significance of Lycopods

(c) Anatomy of stem in *Lycopodium* species

SELAGINELLA

Lycopodiopsida (after Schuettpelz et al., 2016)

[LYCOPODS] (Christenhusz & Chase, 2014)

CLASS LYCOPODIOPSIDA Bartl. (**PPG 1, 2016**)

SUBCLASS LYCOPODIIDAE BEK. (Chase & Reveal, 2009; Christenhusz & Chase, 2014)

ORDER SELAGINELLALES Prantl (Christenhusz & Chase, 2014; **PPG 1, 2016**)

FAMILY SELAGINELLACEAE (Christenhusz & Chase, 2014; **PPG 1, 2016**)

SELAGINELLACEAE (SPIKE MOSSES)

Spike mosses are among the few surviving members of the lycophytes (club mosses, and spike mosses, and quillworts), an ancient group of plants whose origin can be traced back as far as about 400 million years ago. Selaginellaceae within the rank Lycopodiophyta, comprise the oldest lineage of vascular plants surviving on earth (Banks, 2009) with innovations like heterospory and seed habit. Molecular based research indicates that Selaginellaceae also has a close relationship to Lycopodiaceae and Isoetaceae (Tryon and Tryon, 1982), which is chemotaxonomically indicated by the flavonoid patterns (Voirin and Jay, 1978) and it has diverged from Isoetaceae during Upper Devonian period (370 mya) (Korall et al., 1999).

SELAGINELLA

Selaginella, also known as spike moss, is the only surviving genus within the Selaginellaceae family with more than 700 species (about 750, Tryon and Tryon, 1982; Jermy, 1990) widely distributed around the globe (Little et al., 2007). As one of the few lycophyte genera that survived the Permian–Triassic extinction event, *Selaginella* has been a longstanding subject of investigation for botanists and paleontologists. *Selaginella* is estimated congeneric (of same genus) with genus *Selaginellites* Zeller which nowadays remains to fossil (Fairon-Demaret, 1989).

The word *Selaginella* was introduced by Palisot de Beauvois (1805) as a name of a genus. However, previously this word had been used by Linnaeus (1754) for *Lycopodium selaginoides* L.; where selaginoides meant like selago, an ancient name for *Lycopodium*. Subsequently, the name of this species has been revised to *Selaginella selaginoides* (Heidel & Handley, 2006).

TAXONOMY

Although the broad parameters of *Selaginella* are well understood and generally accepted, there have been several attempts to subdivide the genus into genera or subgenera, especially as based on the distinction between the isophyllous, radially symmetrical species and the anisophyllous, bilaterally symmetric ones. Strobilus morphology also varies in their sporophyll shapes; they may be all alike, or of two kinds (dimorphous). The bilateral strobili or the strobili on dorsiventral species may be either resupinate (upside down), with their larger sporophylls in the same plane as the smaller median vegetative leaves, or non-resupinate, with them in the same plane as the larger lateral vegetative leaves.

The simplest division of the genus favoured by Spring (1850), followed by Hieronymus (1901), was by placing the few isophyllous species in subgenus Homeophyllum and the large majority of anisophyllous species in subgenus Heterophyllum. Braun's (1865) division was the same except that he named the subgenera as - Homeotropae and Dichotropae respectively. Walton and Alston (1938) recognised four subgenera with two subgenera of isophyllous plants being *Selaginella* (of Baker) and *Euselaginella* (of Walton & Alston) corresponding to the Homeophyllum of Spring (1850) and Hieronymus (1901) and the Homotropae of Braun (1865), while their other three subgenera of anisophyllous plants *Stachygynandrum*, *Homostachys* and *Heterostachys* together correspond to the subgenera Heterophyllum of Spring (1850) and Hieronymus (1901) and the Dichotropae of Braun (1865). The correlations between these classifications are summarised in Table 4.1. Jermy (1986) improved on this classification by sub-dividing the genus further into five subgenera based on leaf morphology, sporophyll morphology and growth habit. Three of his subgenera (*Selaginella, Ericetorum* and *Tetragonostachys*) are for isophyllous plants and two (*Stachygynandrum* and *Heterostachys*) are for anisophyllous plants. Thomas & Quansah (1991) took the fact that both isophyllous and anisophyllous Selaginella-like plants are known from the late Carboniferous period onwards as supporting evidence that the genus should therefore have a major division into at least two taxonomic units.

From amongst living genera, *Selaginella rupestris*, *S. selaginoides* and *S. pygmaea* are examples of homoeophyllous and *S. kraussiana, S. lepidophylla* and *S. martensii* are heterophyllous.

DISTRIBUTION

Selaginella P. Beauv. - The largest genus of lycophytes/lycopods, with more than 700 species distributed all over the world, is a cosmopolitan genus that grows in tropical and temperate regions. However, highest diversity is seen in the tropical rainforests (Jermy, 1986; Zhou et al. 2015; PPG 1; 2016; Weststrand & Korall, 2016).

Some species of *Selaginella* have a wide distribution and tend to be invasive, but the others are endemics or according to IUCN criteria, they may be endangered.

The Himalayas form one of the diversity centres of ferns and lycophytes of the World. Alston (1945) published the first account of Indian spike moss (including Nepalese) *Selaginella* P. Beauv. Thapa (2002) and later Fraser-Jenkins et al. (2015) listed 23 species of *Selaginella*.

Selaginella grows in the most extreme habitats such as in cold tundra and alpine (*S. selaginoides, S. rupestris*) or in drought dessert (*S. lepidophylla, S. sartorii*); however most of them grow luxuriantly in tropical rain forest.

HABIT

Some species are prostrate, creeping on the ground or over logs or stones, some smaller ones are sub erect or scramble or climb over shrubs with stems several meters long; many are perennial, while a few are small delicate annuals (**PLATE 2A**). *Selaginella oregano* is epiphytic and grows on tree trunks, *S. alligans* is a climber. The xerophytic types *S. lepidophylla* and *S. rupestris*, the northern *Selaginella* sometimes locally known as ledge spike-moss or rock spike-moss, are cespitose. The genus is well adapted to weak light conditions, but at the same time can also thrive in full light conditions of the deserts.

Those species that grow in low light have minute dimorphic leaves, with bizonoplast (*S. erythropus*) and iridescent schemochromic tydall-blue color (*S. willdenowii, S uncinata*); generate more chlorophyll a than b (*S. willdenowii*); and produce several secondary metabolites, mainly bioflavonoid, as well as trehalose, a molecular stabilization of drought conditions (Box 4.1). This adaptation induces continuation of *Selaginella* through time, and triggers the diversity. Biosystematics of *Selaginella* is used for analyzing data obtained from morphological, ecological, genetic, biochemical studies and other parameters to assess the taxonomic relationships, especially within an evolutionary framework. This ascertainment is not only supported by common taxonomic data, but is also supported by newest biomolecular data, the use of molecular markers.

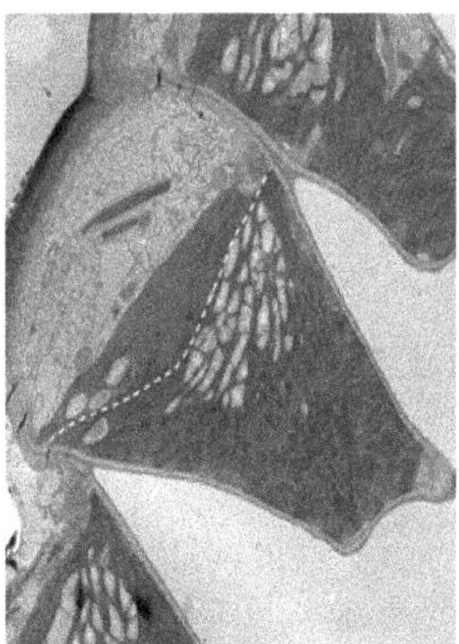

Figure 4.1: Bizonoplast in *Selaginella* species.

Box 4.1

Blue iridescence is most common in the genus *Selaginella. S. willdenowii* (Desv.) Bak., *S. uncinata* Spr. (Iridescent Blue Peacock Fern) and the *S. erythropus* are found in tropical forests. In these species blue iridescence develops on leaves in shade beneath foliage. The green leaves that develop in response to more direct sunlight do not become blue when subjected to this shade, but blue leaves gradually turn green with age or exposure to more direct light. *Selaginella* is able to continue through time exhibiting diversity. Adaptations like bizonoplast (Fig. 4.1) in epidermal cells (**PLATE 2B**), more chlorophyll a than b and production of several secondary metabolites, mainly bioflavonoid, as well as trehalose, enable inhabiting diverse habitats.

Selaginella erythropus, is an understorey shade plant with normal chloroplasts and giant chloroplasts or bizonoplasts both present in individual microphylls, the small leaves (Sheue et al., 2007; Reshak & Sheue, 2012). While biozonoplasts are present in the dorsal epidermal cells, normal chloroplasts are found in the mesophyll and ventral epidermal cells. Sheue et al. (2007) described the ultrastructure of the giant chloroplasts in *S. erythropus* and detected two different kinds of thylakoid arrangements: (a) a lower zone where thylakoids are organized in grana and stroma thylakoids similar to the structure of typical chloroplasts of higher plants and (b) an upper zone that forms a lens-shaped cover of several stacks of three elongated thylakoids – **'bizonoplast'** (BZ). The biozonoplasts fill the majority of the volume of the epidermal cells. The lower region, away from the microphyll surface, demonstrates a standard arrangement of grana (Fig. 4.1); however, the upper region, closest to the microphyll surface, has a highly ordered structure similar to the periodic photonic crystal organization of thylakoid membranes originally seen in *Begonia*. As light interacts with these structures, certain bandwidths of the electromagnetic spectrum are either reflected or transmitted. These structures are responsible for the iridescent blue coloration seen on the leaves of deep-shade adapted understorey plants. It is this structure that results in an increased reflection in the blue region of the visible spectrum. Thus, the chloroplast size reflects an adaptation to increase light capture in the deep-shade environments of the tropical understorey. Plants adapted to deep-shade have to cope also with specks of intense light where the sun breaks through bare areas of the canopy. Sudden exposure to excess light leads to the formation of reactive oxygen species (ROS), which causes oxidative damage to proteins, DNA, and lipids and leads to irreversible damage and necrosis.

Table 4.1: A comparison of the major subdivisions of *Selaginella* into sub-genera.

Reference	Isophyllous species	Anisophyllous species	
		Heterophyllum	Dichotropae
Spring, 1850	*Homeophyllum*	*Stachygynandrum*	*Heterostachys*
Hieronymus, 1901	*Homeotropae*	*Homostachys*	*Heterostachys*
Braun, 1865	*Selaginella*	*Stachygynandrum*	*Heterostachys*
Baker, 1883, 1887	*Euselaginella*	*Homostachys*	
Walton & Alston, 1938	*Selaginella*	*Stachygynandrum*	
	Ericetorum		
Jermy, 1986	*Tetragonostachys*		

SPOROPHYTE

Morphology

Selaginella, a moss-like plant, is similar in appearance to the club moss (e.g., *Lycopodium*). However, the two groups differ as the tiny leaves of spikemoss have a ligule (scale-like flap) at the base of their lower surface (Ligulopsida) and that they are heterosporous - produce two types of spores. *Lycopodium* is without ligule and has been placed under Eligulopsida by early workers. The majority of *Selaginella* species are aniso- or heterophyllous and are placed generally as prostrate with short erect branches. In homoeophyllous forms, the plants are generally erect and leaves that are spirally arranged, are of same size.

- All forms branch freely chiefly in one plane and the forking is dichotomous or pseudomonopodial.
- In species with ascending or sub-erect stems, prop-like structures called rhizophores grow downwards from the forking regions of the stem to the ground giving rise to roots at their tips. These rhizophores provide additional root supply and are believed to support the weak stems.
- The ascending types have probably been derived from creeping types. In heterophyllous (anisophyllous) species with dorsiventral stems, leaves are simple, sessile, small and are arranged in spirals, decussate pairs, or in four longitudinal rows (Fig. 4.2A, B). In most species the branches are with two rows of sublateral or ventral large leaves and two rows of smaller leaves on the upper or the dorsal surface (**PLATE 2C,D**). In certain other species, all the leaves of the four rows are similar and these species are therefore isophyllous (Fig. 4.2C, D).
- The leaves differ from those of Lycopodiaceae in possession of a ligule, a small membranous, tongue-like projection sunken in a pit on the ventral surface near the base.

A **photonic crystal** is a periodic optical nanostructure that affects the motion of photons in much the same way that ionic lattices affect electrons in solids. Photonic crystals occur in nature in the form of structural coloration and animal reflectors, and, in different forms, promise to be useful in a range of applications as seen in wings of butterfly or peacock wing.

- The first root is short-lived, rest arise adventitiously from the undersurface of the stem and from the tips of rhizophores, and remain distributed over much of the plant surface.

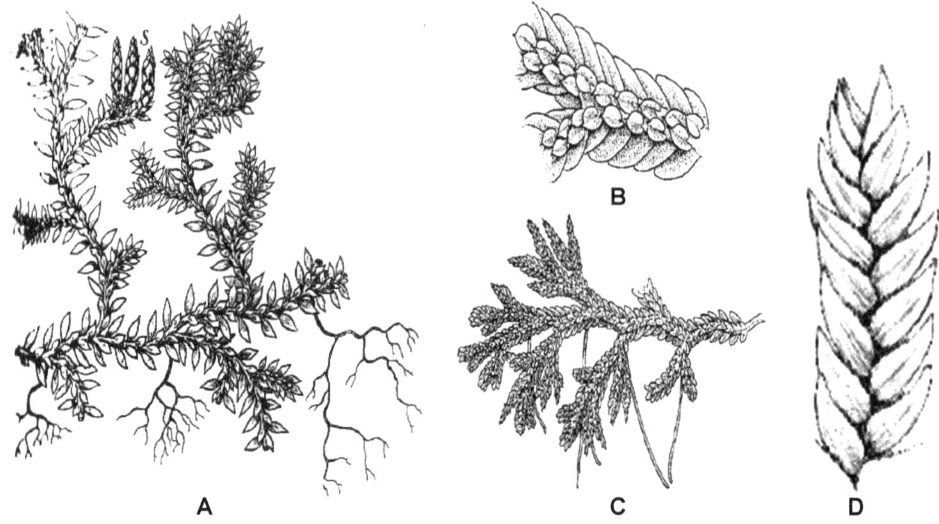

Figure 4.2A-D: *Selaginella* (Morphology)

Diagrammatic construction of anisophyllous and isophyllous species of *Selaginella*.

A – Specimen of anisophyllous species. The plant is dichotomously branched and the rhizophore originates on the ventral side of the branch. B - Close-up of upper side of vegetative branch shows dimorphic leaves, decussately arranged in four rows. C - Isophyllous species. D - Close up of vegetative branch shows uniform spirally arranged leaves.

- The ligule is a flap-like structure on the adaxial surface of the microphylls or leaves of *Selaginella.* Ligules are inserted into a ligular pit, produce callose, achieve maturity before their corresponding leaves, and lack chlorophyll, starch, and intercellular spaces. At maturity, the ligules of each sub genus consist of four sections - these are the sheath, glossopodium or foot, bulbous base, and tip or neck. *Selaginella* ligules are rarely if ever triangular and instead are shaped like a slightly curved, cupped hand with a simple glossopodium (Fig. 4.3A-C). The ligule has long been a functionally enigmatic structure. The tip cells of the ligule contain granular slime. The ligule develops early, is prominent in young leaves but is inconspicuous and shrunken in mature leaves. It is also seen in micro- and megasporophylls (Fig 4.3B). It is suggested that ligule may have a water secreting or water absorbing function to keep young leaves moist.

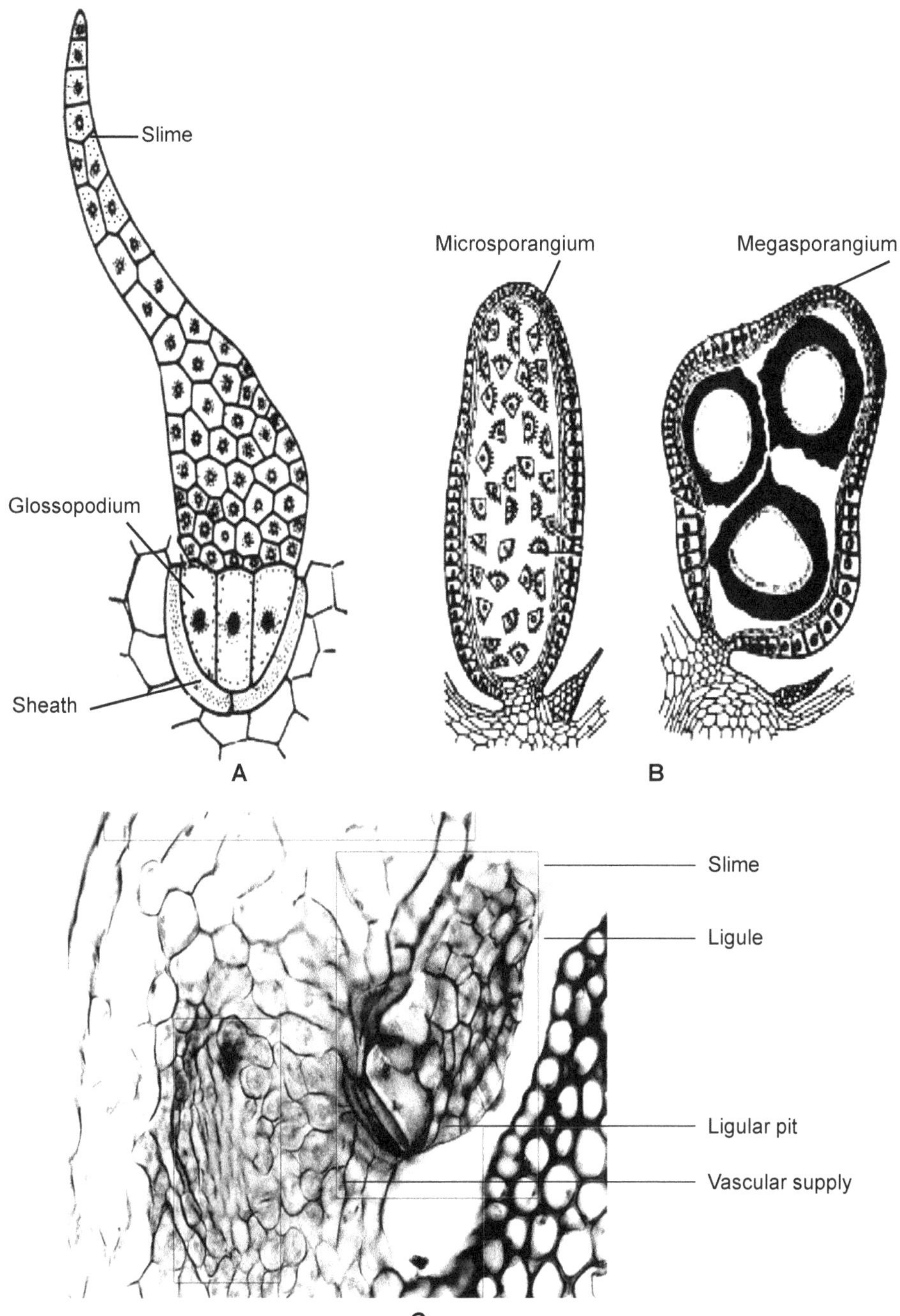

Figure 4.3A-C: ***Selaginella*** **(Ligule)**

A - Section to show glossopodium deeply embedded in ligular pit, and the curved tip. B - The ligule is also present on both, the micro- and megasporophylls. The basal part enlarged to show that the ligule remains connected to the vascular supply of the stem (C).

Anatomy

Stem

- The epidermis consists of cells with outer cutinized walls. Stomata are lacking.
- Cortex in herbaceous species may remain undifferentiated and composed of thin walled parenchyma rich in chloroplasts. In several species with rigid stems, the cortex is differentiated with a several celled thick tissue that comprises hypodermis. This merges gradually with thin walled chlorophyllous cells of inner cortex.
- The stele which remains connected to the cortex through trabeculae, may vary in nature in different regions of the stem. The trailing or prostrate stems and rhizomes have protostele, with two lateral protoxylem strands at two margins. In a single stele (monostele), xylem may become polyarch as in *S. spinulosa* or the xylem may become endarch instead of exarch while the stems may become tristelic or polystelic instead of monostelic (**PLATE 2E, F**).
- In species with radial type of growth habit, the haplostelic condition is characteristic of the lower portion while the upper portion of the stem may be an actinostelic organization or even become siphonostelic if the centre of the stele bears parenchymatous pith. The ribbon-shaped protostele in the rhizome of dorsiventral species may be replaced in the upright branches by a number of concentric bundles (meristeles as seen in *S. wildenovii*); each stele may be regarded as a distinct stele and stems with this condition are termed polystelic.
- Rhizome is solenostelic but upright branches may have 10-15 separate meristeles. The composite vascular cylinder or each meristele is supported in a large air space system by radially elongated endodermal cells designated trabeculae. These trabeculae in most cases represent stretched endodermal cells (Fig. 4.4A).
- Irrespective of stelar organization, the primary xylem is exarch in development (Fig. 4.4B), and the metaxylem consists primarily of tracheids with scalariform pitting.
- In *S. kraussiana*, transverse section shows two steles, where each one is protostelic in organization with phloem placed inner to pericycle. The phloem is formed of phloem parenchyma and sieve cells. The protoxylem has annular and helical thickenings, while metaxylem has scalariform.
- In *S. rupestris, S. oregana*, vessels are present in xylem. In *S. selaginoides,* a few secondary xylem elements have been reported. In this species, the cortex of the basal swelling undergoes a small amount of secondary growth. Some of the secondarily added cortical cells are derived from the periphery of the root-producing meristem, but no cambial layer or layers are ever established.

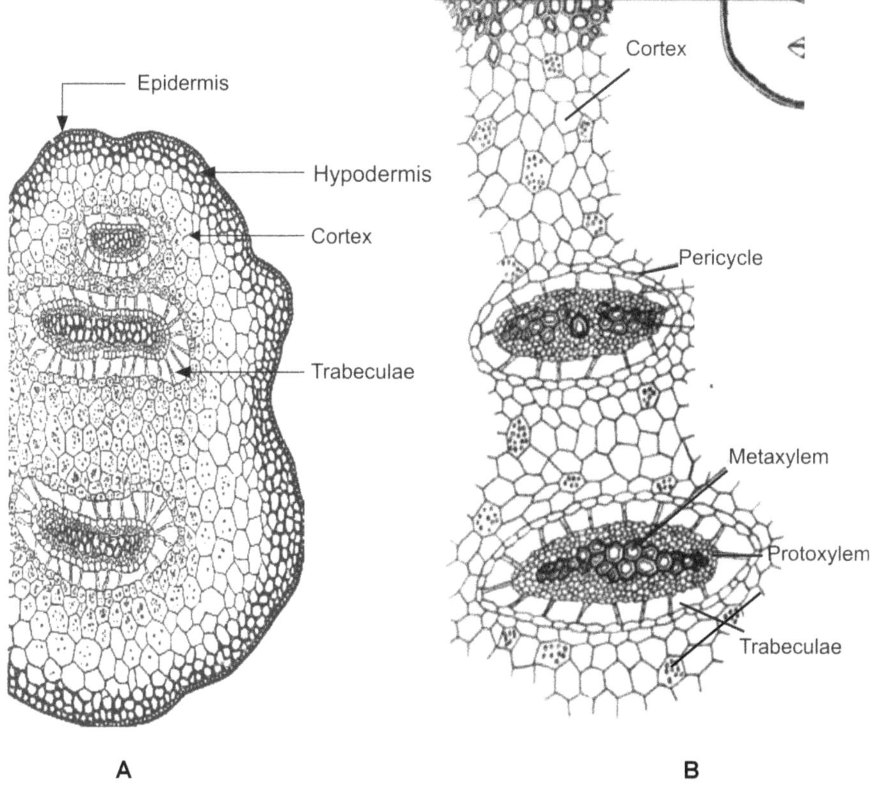

Figure 4.4A, B: *Selaginella* (Stem)

A - Transverse sections to show a tristelic condition. The epidermis is followed by a well differentiated cortex with outer layers modified into hypodermis, which is followed by several layers of parenchyma. B - Steles showing a protostelic organization are magnified.

Leaf

- There is upper and lower epidermis, inner is mesophyll with palisade and spongy layers as in *S. layallii* while in most of the species mesophyll remains undifferentiated. Stomata occur on abaxial surface. There is a median concentric collateral vascular bundle (Fig. 4.5A, **PLATE 2G**).
- Of considerable interest is the large size attained by chloroplasts in this genus. All cells contain a chloroplast. Depending on the species, each mesophyll cell may have one large cup shaped chloroplast or there may be more as many as eight as in *S. willdenovii*. Each chloroplast has a pyrenoid-like structure in the centre.

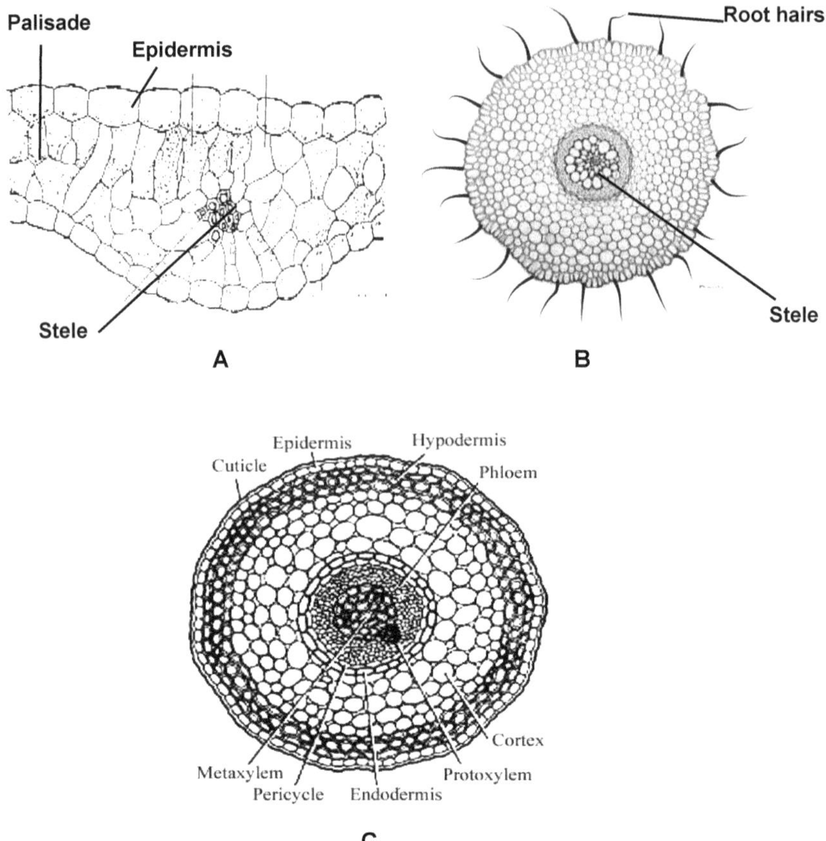

Figure 4.5A-C: *Selaginella* (Anatomy)

A -Vertical section of leaf shows upper and lower epidermis, mesophyll and a median concentric stele. B – Transverse section of root to show piliferous layer with root hairs, a broad cortex and a stele. C - Transverse section of rhizophore to show epidermis, note the absence of root hairs. Also seen is a well defined cortex and a single protostele with monarch xylem.

Root

- Shows epidermis, cortex and stele (Fig. 4.5B). From epidermis arise root hairs. Epidermis is followed by hypodermis and then ordinary parenchyma as in *S. willdenovii* or it may be wholly parenchyma. However in *S. densa,* the entire cortex is sclerenchymatous. The ill-defined endodermis encloses pericycle.
- The stelar region has monarch and exarch condition, with protoxylem situated towards periphery (**PLATE 2H**). In certain species endodermis may be in form of trabeculae as seen in stems.

Rhizophore

As is well known, they are organs that are formed in the branching angle of leafy stems. Cylindrical and of rapid growth, the rhizophores have characters resembling those of roots; their geotropism is positive and they are aphyllous in normal conditions. When they develop above ground, they may give rise to two rhizophores. In a certain species, however, the first ramification usually takes place when the rhizophore makes contact with the soil. This gives rise to roots, organs devoid of chlorophyll, possessing absorbent hairs and which effectively play the role of water absorption.

- Anatomically rhizophore resembles root and has a single protostele. However, the stele may show variations e.g., in *S. martensii* it is monarch exarch while in *S. kraussiana* it is centroxylic with protoxylem in the centre. The cortex and epidermis are sclerified (Fig. 4.5C, **PLATE 2I**).

Nature of rhizophore

Three views exist in literature.

- They are capless roots, they are leafless shoots or they are neither but organs *sui generis*.
- They resemble roots in being positively geotropic, in being leafless and having same anatomy. But they lack cap and root hairs; unlike roots they possess chlorophyll, and have a different anatomy. Certain of these characters alter during the ramification of these organs.
- It resembles stems in having exogenous origin, lacking root hairs and root caps. Under experimental conditions rhizophores develop into leafy shoots. In *S. martensii*, if shoot is injured, incipient rhizophore develops into leafy shoot.
- However, it has been stated by Schoute that rhizophore cannot be *organ sui generis*, as it shares characters with both root and shoot. He further stated that rhizophores are specialized stems but as they are root bearing structures, they are modified in root direction.

Reproduction

Vegetative reproduction

In *S. chrysocaulos* there occur bud-like structures at the tips of some of the vegetative branches that develop underground (Fig. 4.6A,B); while in *S. chrysorrhizos,* the stem apices forming the "buds" repeatedly fork, and rhizophores often occur in the fork between two branches. The behavior of both these reproductive structures, differs since in one of the species the "tubers" remain at the surface of the ground; while in the other they are developed underground, at the ends of filamentous vegetative branches.

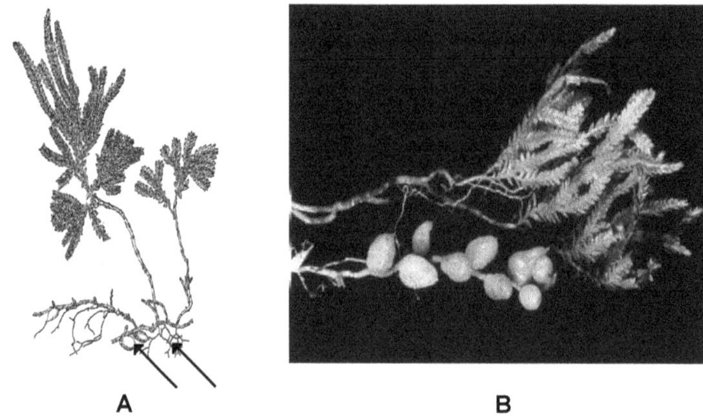

Figure 4.6A, B: *Selaginella moratii* bearing tubers

A - Rhizome with two old tubers (arrows). B -The tubers may be borne in a chain-like fashion.

Salient features of reproduction by spores:

- The life cycle of *Selaginella* involves heterospory, the condition wherein two kinds of spores, microspores and megaspores, are produced in separate sporangia borne in terminal cones or strobili.
- Once shed onto soil, microspores germinate to produce minute endosporic microgametophytes containing antherozoids; and megaspores develop into endosporic megagametophytes containing eggs.
- Once released from the microgametophyte, the antherozoids swim to the eggs in the megagametophyte where fertilization occurs, thus initiating a new sporophyte generation.
- Presumably the separation of sexes on different gametophytes in *Selaginella* results in greater opportunities for outcrossing and increased genetic variation.
- The heterospory is a significant innovation amongst pteridophytes as it lays the foundation of seed habit, seen in gymno- and angiosperms.

Reproduction by spores: Strobili and sporangia

- In most species the fertile region is well differentiated into definite strobili that usually taper towards the apex and vary in size and sex distribution on the axis according to species (**PLATE 2J**).
- Strobili or cones as they are occasionally referred to are usually borne on the tips of main stems or the laterals.
- However, in *S. patuta* and *S. cuspidate*, the axis bearing strobilus may renew its vegetative growth beyond it, terminating in a vegetative shoot. Or as in *S. erythropus* a second strobilus is produced on a fertile branch after an intervening sterile region, thus two fertile regions occur on the same branch.

- The strobilus may be erect or horizontal, rarely pendent. Where it is erect, sporophylls are all alike and are spirally arranged about the axis e.g., *S. selaginoides*. Where the strobilus is horizontal or pendent, it is usually dorsiventral and sporophylls may all be alike or there may be two kinds, the dorsal and ventral ones of different sizes as seen in *S. elegantissima*.
- In most species, the strobili bear both megasporophylls and microsporophylls e.g., *S kruassiana* and *S. helvetia* (**PLATE 2K, L**).
- But in some species e.g., *S. gracilis*, the strobili may possess only one type of sporangia and therefore consist wholly of one type of sporophyll - either megasporophylls or microsporophylls (**PLATE 2M, N**).

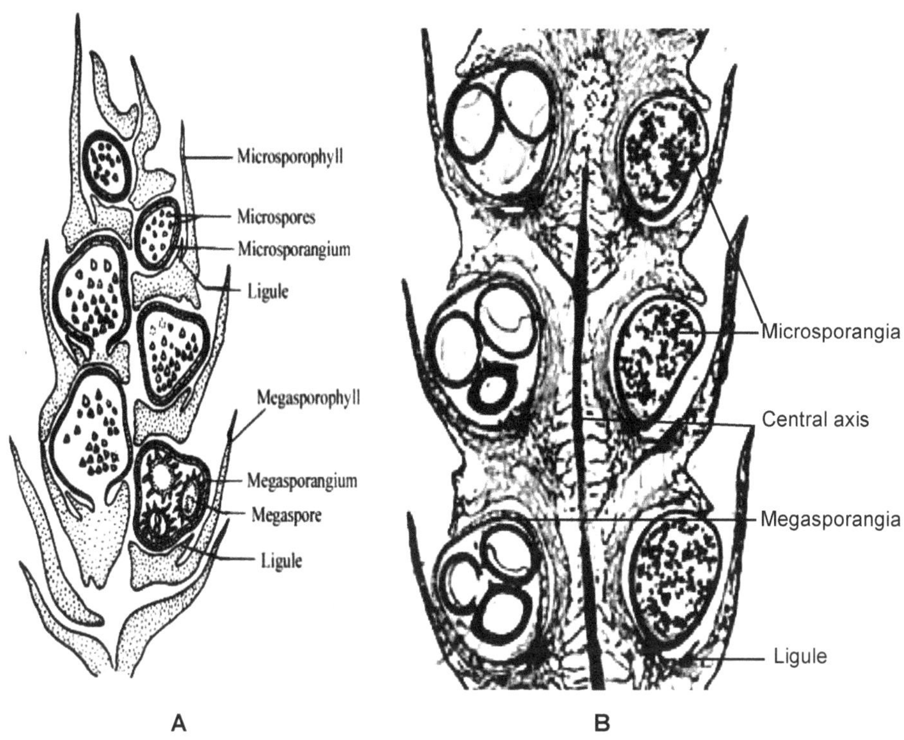

Figure 4.7A,B: *Selaginella* (Strobilus)

A, B - Longitudinal sections to show the tapering tip, central axis and sporangia positioned on the sporophylls. The portion magnified to show presence of a single megasporangium towards the base while rest all are microsporangia (A). In B one half of the longi axis bears micro- and the other half bears the megasporangia. Also seen is the ligule.

- In species where both types of sporophylls occur, their disposition on the axis of the strobilus differs (Fig. 4.7A, B). The strobilus may mostly bear microsporophylls with a single measporophyll at the base as in *S. kraussiana*.

Or there may be two vertical rows of sporophylls one on each side of the axis. One side of the strobilus bears, only microsporophylls and the other side bears only the megasporophylls as in *S. oregano* (**PLATE 2O**). More commonly however, megasporophylls are located regularly in the lower portion of the strobilus and the microsporophylls in the upper portion e.g., *S. selaginoides* and *S. rupestris*.

- In *S. martensii* there is an indiscriminate mingling of micro- and megasporangia.
- The sporophylls also possess a ligule, situated between the end of sporangium and the upturned blade of the sporophyll.
- To sum up one can observe in a Longisection – a central axis of the strobilus around which sporophylls are arranged. The strobili that taper at the distal end, may show variation in distribution of the male and female parts. The microsporangia are generally red in colour and the sporophylls bearing them can be easily observed.

Sporangium

- Each sporophyll bears a sporangium either more nearly upon the adaxial surface or on the surface of the axis, just above the axil of the sporophyll. The sporangia are of two distinct types – megasporangia and microsporangia, the former contain fewer, usually four large megaspores in number but there may be a range as 1, 2, 3, 6, 8, 16, 36 (**PLATE 2P**). The highest number of megaspores is 42 as in *S. willdenovii* and lowest is one, as in *S. monospora* and *S. rupestris*. The microsporangium on the other hand bears many but smaller spores.
- Both micro- and megasporangia are stalked structures.
- The wall of adult microsporangium is a two layered structure. The cells of outer layer are columnar and possess chloroplasts, and this layer is thickened at places. The inner layer is uniform, elongated and thin walled. Innermost is tapetum which persists until spores which are reddish in color, are ready for dispersal. (Fig. 4.8A).
- The megaspore of each *S. rupestris* and *S. apus* has three distinct coats, the exospore, mesospore, and endospore. The former originates on the inner face of the spore mother cell membrane, and when first distinguishable is a film of unequal thickness (**PLATE 2Q**). This either directly or indirectly gives rise to the exospore (Fig. 4.8B) A thick layer develops between the exospore and the protoplasmic vesicle, which later separates into two layers, the mesospore and the endospore.

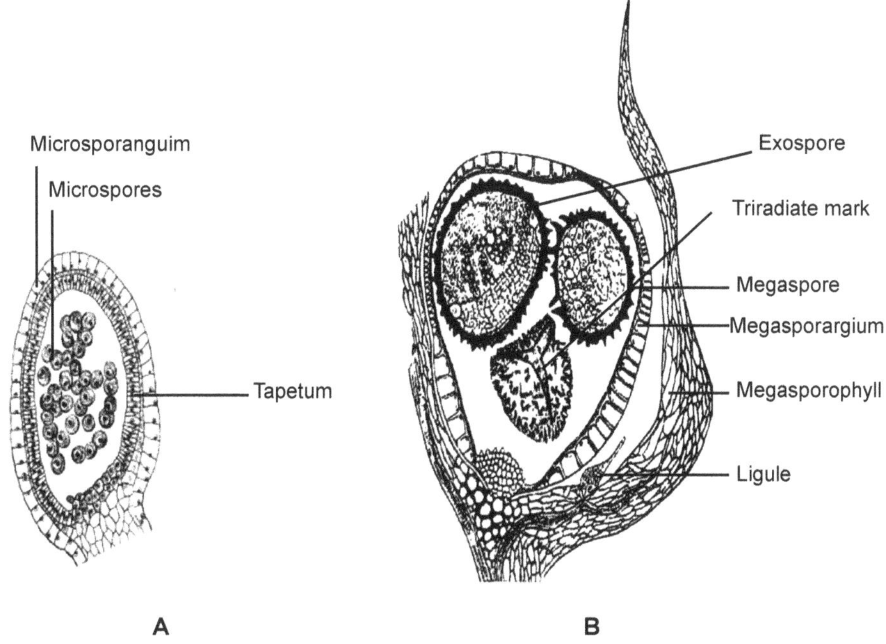

Figure 4.8A, B: *Selaginella* (sporangia)

A-The microsporangium, a stalked structure has a two layered protective wall which encloses a large number of spores. Tapetum, the nutritive tissue differentiates from the outermost layer of sporogenous tissue. B-The megasporangium is a slightly bulged structure and encloses large but only a few megaspores. The megasporangial wall is differentiated into outer and inner wall which appears thickened and has been referred to as endothecium (Rowley & Morbelii, 1995). The innermost is tapetum with a defined role of acting as a conduit for nutrient transfer. The triradiate mark on the megaspore wall is noteworthy.

Dehiscence of Microsporangia

Microsporangia of *Selaginella* have been described as simple ovoid to reniform structures that open at maturity along a distal dehiscence slit, passively discharging microspores from gaping valves (Goebel, 1905; Bierhorst, 1971; Foster & Gifford, 1974). However, studies by Lyon (1901), Mitchell (1910), Gager (1916), and Ingold (1939), indicate that the microsporangia of *Selaginella* do not always conform to simple function but found that in some species they were forcibly discharged. Five anatomically different kinds of microsporangia, which have been divided into three basic types depending on different strategies for microspore dispersal have been described by Koller & Scheckler, (1986). The first type termed the passive type, exhibits passive microspore dispersal, the second type, with two sub-types, shows active microspore ejection and is termed the spore ejector. In the third type termed the sporangium ejector, which also has two sub-types, ejection of the entire microsporangium occurs. All microsporangia at maturity have walls two cell layers thick, where the outer layer shows conspicuous modifications such as the ordering of cells into specific patterns and differential thickening and lignification of cell walls. In both, the spore ejector and sporangium ejector types, specific groups of cells with lignified walls are dead and filled with water when functionally mature. The arrangement of microsporangial stalk cells in cross-sectional view was found to vary little within strobili of one type, but did vary among the types. The passive type dehisces but there does not occur any active ejection mechanism for the microspores. The microsporangium which is reniform in tangential view, and approximately spheroidal in radial view shows a predetermined site for the dehiscence slit. The slit is distally located and extends down the sides along the tangential plane. A mature microsporangium dries up, causing a contraction of the outer layer of cells sufficient to split abaxial and adaxial valves apart, thus initiating dehiscence (Fig. 4.9A-C). Separation of the valves begins at the distal tip and propagates laterally down both sides. The adaxial valve is larger and more cup-like while the abaxial valve is smaller and flatter. The two valves remain attached at their bases but continue to dry up, and their margins curl outward. During this process, the microsporophyll bends outward as well. The dry microspores are now exposed, they separate, and expand into a larger volume. Some microspores fall out over the sides of the microsporangium which either sift down through the strobilus or are carried away on air currents. No further movement by the microsporangium occurs and the remaining microspores are therefore dispersed by wind or water. The central cells in the valves contract as they dry, reducing the size of each valve, and the anticlinal walls become sinuate. Cells surrounding the stalk on the lateral sides can act together like a hinge on which the abaxial valve bends outward as the cells shrink. Lignification of cell walls is greatest along the margins and near the base. There is a slight amount of cell wall thickening and lignification around the stalk.

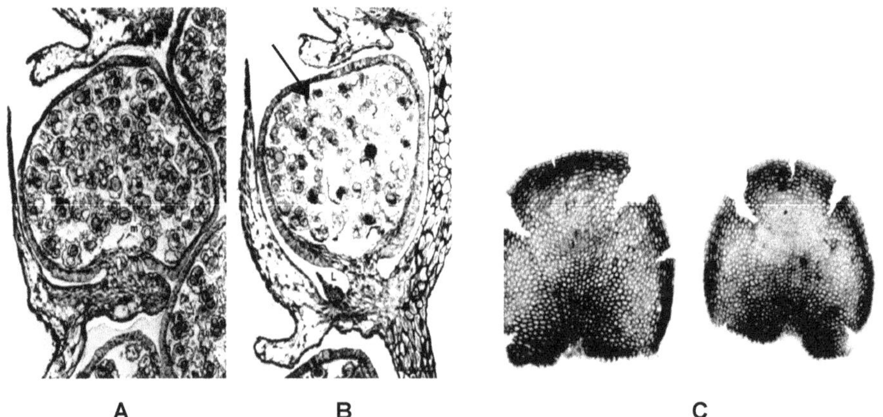

Figure 4.9A-C: *Selaginella* (Microsporangium dehiscence- Passive type)

***S. lepidophylla*. A- Microsporangium is reniform in tangential section and spheroidal in radial section. Dehiscence slit is at arrow (B). C- Microsporangial wall near base. Separation of the valves begins at the distal tip and propagates laterally down both sides.**

Spore ejector type- In this type microspores are actively ejected from microsporangium during dehiscence. The species falling under this type are anisophyllous, and tropical or sub-tropical. Anatomically two different kinds of microsporangia exist, the central sub-type and the basal sub-type. This microsporangium in central sub-type is ovoid in radial and tangential views and it is so called because at maturity, there is an oval region of dead, water-filled cells in the centre of each valve (Fig. 4.10A-C). The anticlinal and inner periclinal walls are thickened and greatly lignified. Two equal valves, abaxial and adaxial, separate along a pre-determined dehiscence slit, but remain attached at their bases. The cells along the valve margins and around the stalk are alive at maturity. The basal sub-type has microsporangium which is spheroidal in radial view, and spheroidal with a truncate base in tangential view. Two valves, abaxial and adaxial are equal, unlike the previous type, and separate along a pre-determined dehiscence slit remaining attached at their bases. The reniform-shaped region of dead, water- filled cells in the basal half of each valve gives it the name of basal sub-type (Fig. 4.11 A-C). The cells have sinuate anticlinal walls, and the anticlinal and inner periclinal walls are thickened and greatly lignified. The cells in the upper half of each valve, and along the margins and around the stalk, are generally alive at maturity. The mechanics of dehiscence and microspore ejection appear similar in both sub-types. When a mature microsporangium dries sufficiently, the two valves begin separating at the distal tip and the dehiscence slit propagates laterally down the sides accompanied by immediate reflexing of the margins. This is more pronounced in the central sub-type (Fig. 4.10C). The valves begin to bend away from each other. This bending is localized in the central region of each valve of the central sub-type, and in the basal region of each valve of the basal sub- type. Once exposed, the microspores dry and separate into a larger volume while the valves continue to reflex and bend away from each other. The valves then snap back toward their original positions ejecting a large portion of the microspores and immediately begin to reflex again, and again they snap back. This occurs up to about seven consecutive times before the movement stops, at which time all the microspores are normally ejected. In a dried strobilus of either of these sub-types, plumes of ejected microspores may be seen around it. After dehiscence is complete and the valves stop the movement, all water-filled cells now contain gas bubbles which dissolve when microsporangia are soaked in water, after which a complete sequence of opening and snapping happens again upon drying.

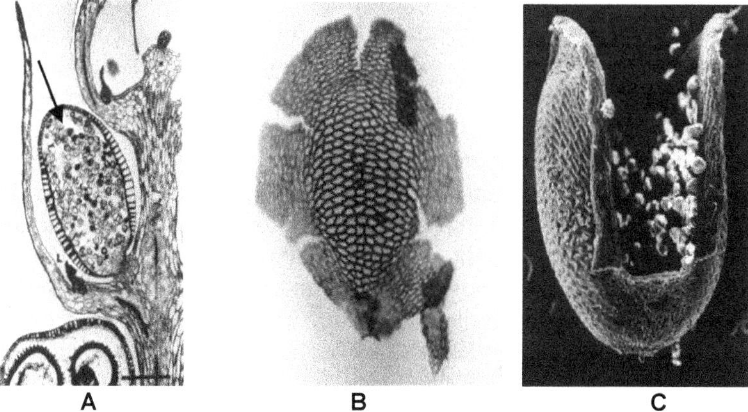

Figure 4.10 A-C: *Selaginella* (Microsporangium dehiscence-spore ejector type)

Spore ejector type (Central sub-type) in *S. emmeliana*. A- Microsporangium is oval in radial section. Dehiscence slit is at arrow. B -Two valves separate along line of dehiscence but remain attached at the base. C- One valve oval in shape in tangential view, and lignified cells are present in central region.

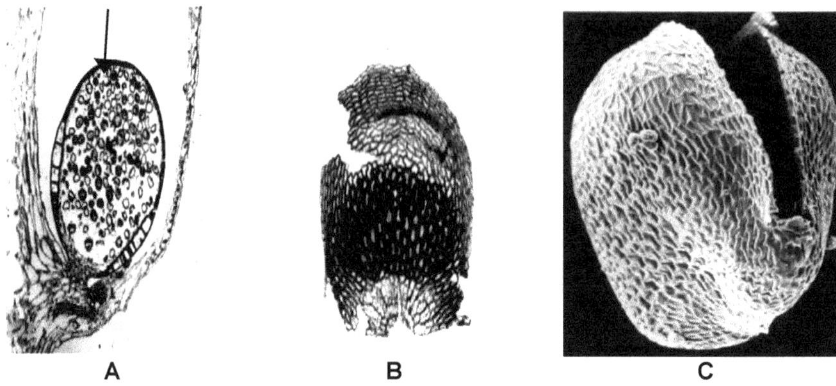

Figure 4.11A -C: *S. plana* (Spore ejector type-Basal sub-type)

A–Microsporangium is spheroidal in radial section. Dehiscence slit is at arrow. B- Both the valves formed along the line of dehiscence are equal. C- One valve to show lignified cells in basal region of valve.

Sporangium ejector-The microsporangium is ejected from the strobilus during dehiscence. All species in this type are anisophyllous and tropical. The microsporangia from all these species are essentially identical, but in *S. kraussiana* some distinct anatomical differences exist. Thus, there are two sub-types of sporangium ejector. The microsporangium is spheroidal in radial section, and ovoid in tangential section. On the external surface is a raised band comprised of large cells of the outer microsporangium wall that runs radially around, splitting into two bands around the stalk. The band is interrupted on the adaxial surface which is the distal point of the dehiscence slit. As a result, during dehiscence an abaxial valve is produced that is larger than the adaxial valve. In addition, the two complete lateral sides of the microsporangium remain attached to the ends of the abaxial valve. In *S. kraussiana*, the raised band of cells is interrupted at the top of the microsporangium rather than on the adaxial face. As a result two equal valves are produced. The two lateral sides of the microsporangium are equally divided into four flaps, two of which remain attached to each valve. Cells along the margins and around the stalk are alive at maturity, while the cells comprising the raised band are dead and water-filled with thickened and lignified anticlinal and inner periclinal walls. When a mature microsporangium dries sufficiently the two valves separate, at the point where the raised band is interrupted on the adaxial face. The large abaxial valve moves up and bends outward and gets exerted over the subtending microsporophyll, except in *S. kraussiana*. The microsporophyll is pushed outward in the process (Fig. 4.12A-C). The smaller adaxial valve reflexes, and pushes against the strobilus axis. The open microsporangium carrying most of the microspores is suddenly ejected from the strobilus. The microsporophyll snaps back to its original position. The direction of microsporangium ejection is approximately that in which the strobilus is pointing. Microsporangia are ejected up to 20.5 cm away with a large number of microspores still clustered within. The valves of the microsporangium immediately continue reflexing, and may fold completely over one another. Occasionally the microsporangium may stay in this position, with usually a series of small snapping motions, releasing the microsporangium back to a more open position. The snapping motions bring about microspore dispersal. When movement stops the microsporangium usually remains in an open position.

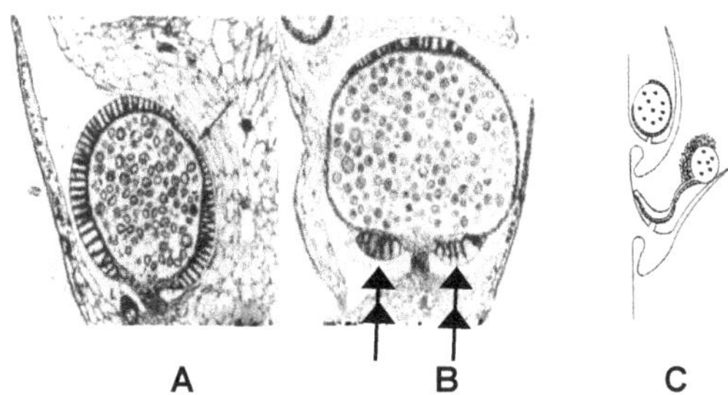

Figure 12A-C *Selaginella* (Microsporangium dehiscence-Sporangium ejector type)

A–Microsporangium is spheroidal in radial section. Dehiscence slit is on adaxial face (arrow). Raised band of cells (double arrows) forms the outer wall of the microsporangium. B-The raised band extends over the top and splits in two around the stalk. C- The smaller adaxial valve bends back against the strobilus axis. The larger abaxial valve, with lateral sides of the microsporangium attached, reflexes against the subtending microsporophyll. Dots represent microspores.

Dehiscence of Megasporangium

Since the male and female sporangia vary in structure and the number of spores produced, it would be worthwhile to know the number of spores produced by the megasporangium. There is a great variation and examples of increase in megaspore numbers above four is seen in *S. willdenowii, S. inaequalifolia v. perelegansy, S. lobbii, S. watsoniana*, and *S. serpens*. Reductions in megaspore numbers below four is encountered in *S. willdenowii, S. inaequalifolia v. perelegans, S. erythropus, S. bakeriana S. stenophylla* and *S. chrysorrhizos*. In *S. martensii* which also produces four megaspores, the megasporangium has distinct bulges at places where the four large white megaspores are contained inside.

Certain regions of the megasporangium such as hinge, annular patch, and distal line take part in dehiscence and their role is discussed. The annular patch is restricted to the basal region of the megasporangium, while the line of dehiscence is distal as seen in microsporangium.

The megasporangium has a region of translucent thin-walled cells that extends along the base and forms the conspicuous "hinge". The hinge plays an important role in spore dispersal.

Once the megasporangium has reached maturity, a split occurs at the distal end and proceeds along the line of dehiscence down the sides of the sporangium, resulting in formation of two flaps and a basal "boat".

The flaps bend backwards as the sporangium opens, exposing the four megaspores. Two basal megaspores lie within the boat, while two apical spores, oriented at right angles to the basal spores, are positioned on the surface of each flap.

As the flaps reflex, the basal thin-walled hinge cells gradually collapse, resulting in narrowing of the boat. The constricting sides of the boat exert pressure on the two basal spores, squeezing them slowly upwards. Due to the pressure the total collapse of the basal hinge and boat occurs, resulting is sudden discharge of spores.

With the sudden collapse of the boat, the flaps of the megasporangium are jerked toward one another, and the two spores placed at the top of the boat are rapidly ejected, more or less simultaneously with the spores at the base of the boat (Fig. 4.13A-D). In *S. selaginoides*, it is described as the "compression and slingshot" mechanism. Although actual distances of microspore and megaspore dispersal cited by different investigators vary, it is clear that megaspores are shot to a greater distance. For example microspores of *S. martensii* are shed for a distance of approximately 1 cm and megaspores are propelled 10-20 times further (Ingold, 1974).

**According to the researchers, this kind of dispersal brings closer spores from different plants, an adaptation for cross fertilization. The strobili have protandrous condition with microspores germinating sooner on the same strobilus and archegonia developing six weeks after antheridia are empty.

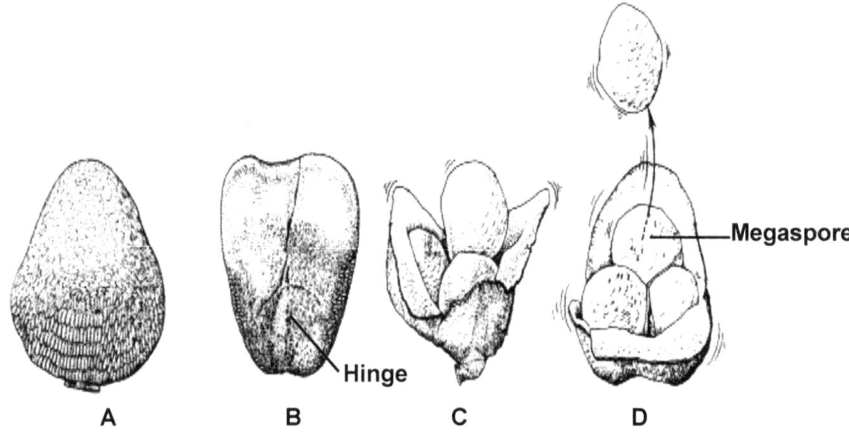

Figure 4.13A-D: *Selaginella* (Megaspore dehiscence)

A–The line of dehiscence is distal while the annular patch is basal. Hinge which plays an important function in dehiscence is placed very close to the annular patch. B,C-The dehiscence proceeds downwards along the line of dehiscence as a result of which two flaps and a boat is formed. D- Out of four megaspores, two are placed in a position from where they can be immediately dispersed when the hinge tissue contracts. The other two placed at the base of the boat are then ejected.

GAMETOPHYTE

Megaspore Germination and Development of Megagametophyte

The development of megagametophyte is *in situ* and begins when the megaspore is still present inside megasporangium. In fact development begins long before megaspore has reached its full size. The stage of development at the time of shedding of megaspore varies from species to species. In some species megaspore is shed after a few divisions have taken place but before megagametophyte becomes cellular. In *S. karaussiana* it is usually shed shortly after the first archegonia are formed. In *S. rupestris*, megaspore may be retained within megasporangium until fertilization has taken place and embryo has started to develop further. But in *S. spinulosa*, development of megagametophyte starts only after megaspore has been liberated from the megasporangium.

At maturity each spore has a trefoil-shaped cleft extending from the apex along the three ridges between the triangular faces, and is bound by the flaring flaps of the torn exospore. The initial step of the female gametophyte development is the rapid expansion of megaspore, accompanied by karyokinesis. The nuclei are positioned peripheral and central vacuole is organized. At this stage, the endospore and mesospore appear as thin but distinct film-like layers. The gametophyte at this stage thus consists of (i) exospore, still growing, (2) meso- and endospore, both thin; (3) the vesicle consisting of a very thin and homogeneous layer of protoplasm applied to the inner surface of the endospore, in whose apical region are imbedded numerous large, ovate, flattened nuclei; and (4) a large central vacuole filled with a watery fluid in which are suspended many oil drops.

When megagametophyte attains maximum size, cell walls are laid down and the protoplasm encroaches on the space occupied by the vacuole which finally gets obliterated. The cells towards the apex or the trefoil shaped cleft are smaller in size and form a prothallial cushion. The rest of the megagametophyte is filled with larger food laden cells (Fig. 4.14A, B).

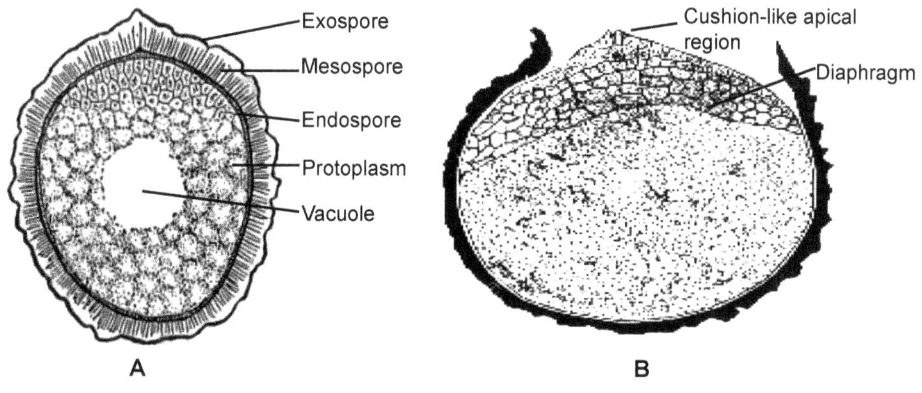

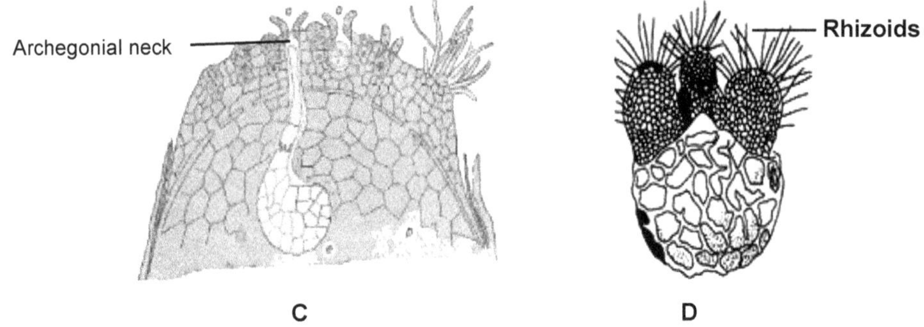

Figure 4.14A-D: *Selaginella* (Megagametophyte)

A - Free nuclear divisions result in multinucleate condition and these nuclei get arranged towards periphery with a central vacuole. B- Cell wall formation then ensues in the apical region forming a cushion and the vacuole gets obliterated. The lowermost layer of cells of the cushion get thickened and a diaphragm is formed. C -Archegonia arise on the surface of the cushion and are embedded structures with only necks protruding out. D- Mounds of tissue covered with rhizoids break open the megaspore.

In some species the lowermost cells of cushion become thickened and form a diaphragm that separates it from the rest of the gametophyte. In *S. rupestris*, there is no diaphragm, and at no stage of its development are there nuclei in the lower or the central portions of the gametophyte, which at first contains liquid, and finally a semisolid mass of granular matter.

Archegonial initials differentiate on the surface of the cushion (Fig. 4.14C). The archegonia are embedded in the gametophyte at the venter region and their two celled necks project at the gametophytic surface (**PLATE 2R**). The axial row comprises one neck canal cell, one venter canal cell and an egg cell. On the cushion, under three arms of triradiate mark, appear mounds of tissue, which remain covered with rhizoids and upon further growth these mounds crack open the megaspore wall (Fig. 4.14D). The mounds are prominent in some species, and inconspicuous in others. The rhizoids help entangle the antherozoids and facilitate fertilization.

Germination of microspore and development of microgametophyte

The microspore germinates *in situ* and generally a 13- celled gametophyte is formed before the microsporangium dehisces. These cells are – 1 prothallial cell+ 8 jacket cells+ 4 androgonial cells. After this stage variation may occur in development depending on the species. In some cases gametophytes may be shed a little earlier while in others they may be shed a little after 13-celled stage by which time considerable development has occurred. However, in most cases, some part of the development occurs in soil. The immature male gametophytes may fall in the clefts of partially opened female gametophytes and further development of male occurs in close proximity to megagametophytes. This is occasionally referred to as 'incipient type of pollination'. Once in proximity of megagametophytes, each of the four androgonial cells may undergo six successive divisions resulting in a total of 256 antherozoid mother cells. The jacket disintegrates and the antherozoid mother cells float free in the slime present inside the cavity. Each mother cell metamorphoses into an antherozoid – a spirally coiled body with two terminal flagella. Thus both *Lycopodium* and *Selaginella* have this bryophyte character of two flagella in contrast to other pteridophytes (ferns) which have more.

Microgametophyte a highly reduced structure, never outgrows the bounds of microspore wall and is never set free. It remains dependent on parent sporophyte for food. The microgametophyte development is considered as merely a case of cell division and cell differentiation in which actual growth processes are largely eliminated. Most workers believe that a single antheridium per microgametophyte exists in *Selaginella*. According to others depending upon species, two or four initials are formed. Hence four antheridia per microgametophyte as in *S. kraussiana* and two antheridia per microgametophye as in *S. martensii* are reported. When antherozoids are mature, swelling of mucilage by absorption of water bursts the wall of microspore along the triradiate ridge and biflagellate antherozoids are set free in film of water.

YOUNG SPOROPHYTE

After fertilization, the zygote secretes a thick wall and develops into an embryo. The first division in the zygote is usually transverse (but it may be vertical as in *S. galeottii*) and two daughter cells are formed. The upper epibasal cell (the cell next to archegonium neck) is called the suspensor cell and the lower (hypobasal cell) is known as the embryonal cell. The embryo is thus endoscopic. The further development of the embryo in which a suspensor consisting of one or several cells, the apex of the stem with the cotyledons, the first rhizophore and the foot are distinguishable, proceeds in a great variety of ways. In *S. kraussiana* and *S. galeottii* the suspensor is much reduced and is replaced by another organ of analogous position, the embryo tube which does not belong to the embryo but develops from the basal archegonial cavity surrounding the egg cell. In this case, the archegonial cavity grows rapidly and penetrates deep into the tissue of the gametophyte thus forming the elongated embryo tube. The developing embryo lies at the inner extremity of the embryo tube.

In *S. denticulate*, the upper (epibasal) cell develops the suspensor, and from the basal cells of the suspensor adjacent to the hypocotyl is formed, a large foot. From the upper part of the foot is formed a conical protuberance, which is the primary rhizophore; the first roots later arise from it. The lower (hypobasal) cell gives rise only to the stem, cotyledon and the hypocotyl.

In *S. martensii* the upper (epibasal) cell of the two celled embryo forms only the suspensor and the lower (hypobasal) cell forms all the other organs of the embryo sporophyte. In this species, the epibasal or suspensor cell grows rapidly down wards and thrusts the embryo proper down into the large nutritive prothallial tissue. The hypobasal or embryonal cell gives rise to foot at right angles to suspensor, two cotyledons, and a ligule. Later a hypocotyledonary part of the stem is formed. Root which is aligned in a manner that it soon reaches the ruptured part of megaspore wall, arises from the superficial cell of the foot. From the apical cell of the root is formed the rhizophore.

The young sporophyte remains attached for some time to the megagametophyte, still held within megaspore wall, later grows upwards. The rhizophore bends down wards and root enters the ground while the stem develops and branches thereafter (Fig. 4.15A-C).

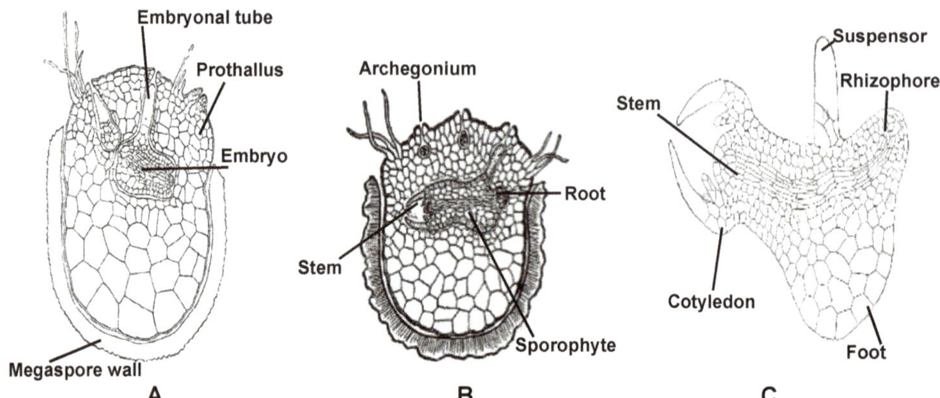

Figure 4.15A-C: *Selaginella* (Stages in development of young sporophyte)

There is distinct alternation of generation in *Selaginella* (Fig 4.16 A-G) and significant stages of heterospory are seen.

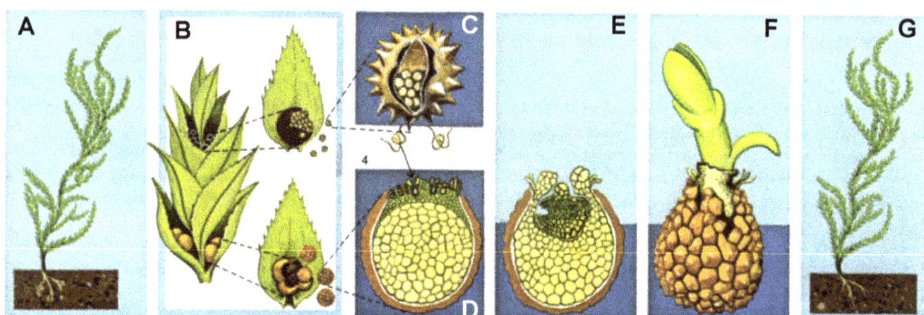

Figure 4.16A-G: *Selaginella* (Life cycle stages)

SOME INTERESTING FACTS

- The spikemoss *Selaginella moellendorffii* has a genome size of only ~100Mbp, which is the smallest genome size of any plant reported. The sequence of the *Selaginella* genome (by JGI) provides scientists an important reference genome necessary for deciphering the evolution of biochemical, physiological and developmental processes unique to land plants. (Source:*The JGI Genome Portal*).

- Although traditionally grouped with other spore-producing plants such as whisk ferns, horsetails and ferns, we now know spike mosses and club mosses are not closely related to other spore producers.
- The dry land species *Selaginella lepidophylla*, a native of the deserts of Chihuahua, Mexico is also known as the resurrection plant due to its ability to come back to life after withstanding years without water. However, following rain it will sprout new green leaves. Spanish missionaries apparently noted this property when they arrived in the New World, and quoted it to illustrate the concept of rebirth to potential converts. *S. bryopteris* or sanjeevani (one that infuses life) has mythological importance in Indian culture.
- *Selaginella willdenowii* is a species with iridescent blue foliage. The colouring is a consequence of elaborate structures on the upper surface of the leaves.
- The earliest documentation of *Selaginella*-based treatments appeared in Shen Nong Ben Cao Jing (The Divine Farmer's Materia Medica) in 2737 BC, where *Selaginella* was used to treat inflammation, amenorrhea, and abdominal lumps in women (Yang & Flaws, 1998). *S. bryopteris* or sanjeevani (one that infuses life) has been used for centuries in Indian ayurvedic medicine to treat burning urination, menstrual irregularities, and jaundice (Sah et al., 2005). Despite a long history of *Selaginella* being used as an herbal remedy and tonic, the scientific basis underpinning its efficacy in treating various maladies is lacking.
- Around 130 natural products, previously reported from *Selaginella*, have been sorted into six of the major categories of plant specialized metabolites including flavonoids, lignans, selaginellins, other phenolics, alkaloids, and terpenoids.
- *Selaginella* has various pigmentations, such as blue chromatic, crimson red, variegate, yellow gold, and silver. Morphological diversity and pigmentation are important characteristic in taxonomy of *Selaginella* (Dahlen, 1988; Czeladzinski, 2003). Shape of stele (Hieronymus, 1901) and structure of sporangium (Horner & Arnott, 1963; Fraile & Lap, 1981; Quansah, 1988) are also used as distinguishing characteristics.
- Major threat to sustainability of *Selaginella* is habitat conversion and fragmentation. Habitat conversion can bring about total loss of *Selaginella* due to microclimatic changes, as a result of water way changes, road-works, husbandry, fire-burning, deforestation, and development of recreation area (Heidel & Handley, 2006). Habitat fragmentation can induce inbreeding depression degrading offspring endurance (Barrington, 1993), which is observed at fragmented spots of *S. selaginoides* communities. However, global warming is the recent threat that can change diversity and sustainability of organism in the longer time period.
- On the plus side of growing Selaginellas in gardens is that it is seldom damaged by pests or diseases. Neither do these demand much feeding, in fact too much fertilizer may kill plants. Re-potting may become necessary as clumps enlarge eventually and these are then suited to shallow trays

rather than deeper containers. Likewise plants can be split and multiplied in trays in early spring. It is possible to grow these from the spores but it may turn out a less successful choice.

Selaginella talk	
Selaginella braunii	Arborvitae Fern, Arborvitae Spikemoss, Braun's Spikemoss, Chinese Lace Fern.
Selaginella kraussiana	Club Moss, Trailing Spikemoss, Golden spike moss
Selaginella lepidophylla	Resurrection Plant
Selaginella pallescens	Dwarf Cedar Fern
Selaginella stauntoniana	Staunton's Spikemoss
Selaginella uncinata	Peacock Moss
Selaginella kraussiana	Krauss' spike moss, Forest selaginella, Spreading spike moss
Selaginella acanthonota	Spiny spikemoss
Selaginella rupestris	Spring-ledge spikemoss, rock spikemoss, northern selaginella
Selaginella densa	Small clubmoss, dense clubmoss, lesser spikemoss

Class LYCOPODIOPSIDA
Order SELAGINELLALES
Family SELAGINELLACEAE
SELAGINELLA

QUESTIONS

Q1. 'Heterospory has led to seed habit in land plants'. Comment on the statement, taking *Selaginella* as an example.

Q2. Elucidate the salient features of life cycle in spike moss.

Q3. With the help of neat diagram give the anatomical features of rhizophore and bring out its organization close to both stem and root.

Q4. The development of embryo in *Selaginella* sp. is endoscopic. Discuss.

Q5. Give the evolutionary significance of *in situ* development of megagametophyte in the spike moss.

POLYPODIOPSIDA

(Schuettpelz et al; 2016)

The ferns constitute an ancient division of vascular plants, some of them as old as the Carboniferous Period (beginning about 358.9 million years ago) and perhaps older with records extending from the middle or late Devonian periods.

Ferns differ profoundly from flowering plants and are defined by several characters which form basis of classification of this group. They are vascular plants that produce spores and undergo an alternation of generations (with separate gametophyte and sporophyte generations that exist as free-living plants). Though mosses, hornworts and liverworts (bryophytes) are also spore-producing plants with alternation of generations, but their physiological dependence on gametophytes limits size and they never attain heights while true ferns can be trees with significant height.

GENERAL FEATURES

Size Range and Habitat

The ferns are extremely diverse in habitat, form, and reproductive methods. They range from minute delicate plants only 1–1.2 cm (0.39–0.47 inch) tall to huge tree ferns reaching a height of 10 to 25 metres (30 to 80 feet). Some are twining and vine-like (*Polypodium aureum*); others are aquatic floating on the surface of ponds (*Salvinia*). The majority of ferns inhabit warm, damp areas and grow luxuriantly in tropical areas. A few are found in dry, cold places. Some species play an important role in ecological succession, and inhabit the crevices of bare rock. They are also the first occupants in open bogs and marshes prior to the advent of forest vegetation. The best-known fern genus found all over the world, *Pteridium* (bracken) is characteristically found in old fields or cleared forests, where it is often succeeded by woody vegetation.

Distribution and Abundance

As mentioned, ferns are most abundant in the tropics while Arctic and Antarctic regions possess only a few species. The tropical rainforests, display huge diversity with a few species constituting a dominant element of the vegetation. Also, many of the species grow as epiphytes on the trunks and branches of trees. A number of families e.g., Marattiaceae, Gleicheniaceae, Schizaeaceae, Cyatheaceae, Blechnaceae, and Davalliaceae are almost exclusively tropical while most of the other families are found in both the tropics and the temperate zones. Only a few genera e.g., *Athyrium*, *Cystopteris*, *Dryopteris*, and *Polystichum* that are primarily found in temperate and Arctic regions, extend into the tropics, where they are found at high elevations on mountain ranges and near volcanoes.

Ferns are uncommon as invasive species outside of their native ranges, although a few such as bracken **(*Pteridium*)**, spread quickly by underground rhizome. One species of water spangles (*Salvinia auriculata*) has become a major pest in India, blocking irrigation ditches and rice paddies. Another species (*S. molesta*) became invasive in southern Africa, killing other plant life and fish by cutting off light and oxygen. Examples of other invasive ferns include the giant polypody (*Microsorum scolopendrium*), climbing ferns (*Lygodium japonicum* and *L. microphyllum*), green cliff brake (*Cheilanthes viridis*), silver fern (*Pityrogramma calomelanos*), Japanese holly fern (*Cyrtomium falcatum*), rosy maidenhair (*Adiantum hispidulum*), Cretan brake (*Pteris cretica*), and ladder brake (*P. vittata*). Two Old World species (*Cyclosorus dentatus* and *Macrothelypteris torresiana*) were introduced into tropical America around 1930 and are now among the most common species even in some remote areas.

Alternation of Generations

The leafy fern plant growing in fields, marshes and forests is a sporophyte, a vascular plant that through meiosis produces spores with half the number of chromosomes found in the mother plant, the sporophyte. On a suitable substratum, it grows into a free-living gametophyte, a haploid plant. Typically gametophytes are photosynthetic, but there are mycoheterotrophic gametophytes in some fern genera such as *Sceptridium dissectum* and *Ophioglossum crotalophoroides.* In addition to having mycoheterotrophic gametophytes, there are a few members of *Botrychium* that are unique among ferns in having the sporophytes also mycoheterotrophic, producing only small, ephemeral sporophylls that do not photosynthesize. Gametophytes, often called prothalli, produce through mitosis, gametes. The male gametes, antherozoids which are motile, free-swimming, flagellate are formed inside antheridia. They swim to the archegonium formed on prothallus and contact the egg located inside the venter. For this act of fertilization which results in zygote, the first stage of sporophytic generation, ferns require water. After fertilization the zygote and the resultant embryo are held within the tissues of the gametophyte. The embryo grows into an independent plant and the prothallia wither and die. Exceptions to this life cycle include several aquatic heterosporous genera with separate megaspores and microspores, and reduced gametophytic phase that remains largely within the spore walls. Several other ferns are apomictic where new sporophytes arise directly from cells of the gametophyte without the act of fertilization.

The gametophytes can be bisexual or unisexual, and some can also reproduce vegetatively (a few may have lost the sporophyte stage altogether). In the life cycle of bryophytes also alteration of generation occurs, but in this group, pattern differs in the sense that the sporophytes are independent while in bryophytes, sporophytes are parasitic on the photosynthetic gametophyte. Ferns also differ from higher vascular plants in the sense that in the latter, male and female gametophytes develop inside the pollen and ovule, respectively, with ovules being retained on the parental sporophyte until after fertilization, and later it is released as a seed with the new sporophyte partly developed inside the ovary wall.

Though the sporophytes reproduce through spores they may also propagate vegetatively by branching of rhizome, often forming large, genetically uniform plants by latter means (Fig. 5.1A). A few ferns propagate by root proliferations, and some, especially in the wet tropics, reproduce by leaf proliferations.

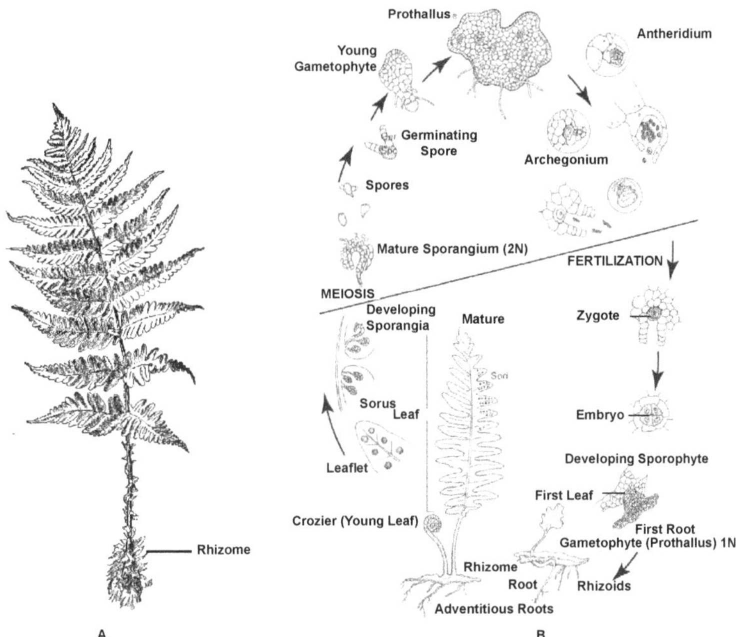

Figure 5.1A,B: (A)Mature frond of *Aspidium marginale*, (B) Life cycle of a fern to show alternation of generations.

Life Cycle

The life cycle of the fern shows (1) sori borne on the undersurface of modified leaves, the sporophylls. (2) upon release, the haploid spore is carried to the ground, where it germinates into a tiny, usually heart-shaped, gametophyte (gamete-producing structure), the prothallus anchored to the ground by rhizoids. (3) under moist conditions, antherozoids are released from the antheridia and swim to the egg-producing archegonia that have formed on the lower surface of the prothallus. (4) upon fertilization, a zygote forms that develops into an embryo within the archegonium. (5) the embryo eventually differentiates into root, shoot and leaves, grows larger than the gametophyte and becomes a sporophyte which is an independent structure (Fig. 5.1B).

Ecology

Ecologically, the ferns are plants of shaded damp forests of both temperate and tropical zones. Some fern species grow equally well on soil and on rocks while others are confined strictly to rocky habitats, occupying fissures and crevices of cliff faces, boulders, and taluses. Acidic rocks such as granites (*Anogramma leptophylla)*, sandstones, and quartzites harbour a characteristic fern flora distinct from that of alkaline rocks such as calcites and dolomites. A few species appear to be confined to serpentine and related rocks. In the tropical habitats a large number of the ferns of an area may be epiphytic on the shaded lower trunks and branches or in the crowns of trees. A few of these epiphytic ferns are climbers (*Lygodium japonicum* and *L. microphyllum* are climbers, and *Asplenium* and *Platycerium* are epiphytic). In these species, the lower leaves (bathyphylls) are usually vegetative and often differ in form from those at the higher levels (acrophylls), which are entirely or partly fertile bearing sporangia.

Both epipetric (growing on rocks) and epiphytic ferns may show structural adaptations to dry habitats such as presence of hard tissue, the epidermis, with a thick cuticle (a waxy layer); abundant trichomes or scales on the leaf and stem. Such modifications are also seen in especially those species that grow in salt marshes (e.g., leather ferns, *Acrostichum*) and in open, fully exposed sites (e.g., bracken, *Pteridium*; lip ferns, *Cheilanthes*; and brakes, *Pteris*). Ferns that grow in the open are often referred to as sun ferns (e.g., *Gleichenia*) and, differ from species that require shade. Water ferns—water clovers (*Marsilea*), water spangles (*Salvinia*), and mosquito ferns (*Azolla*) appear only after rains, and complete their life cycles rapidly, probably as an adaptation to the need for making quick use of water. These ferns are heterosporous and essentially lack the vegetative phase of other ferns. Many inhabitants (Pteridaceae), of dry rocky slopes and cliffs, have a modified apogamous life cycle that bypasses fertilization. This life cycle is also believed to foster quick reproduction in connection with brief damp periods. The gametophytes grow quickly, with buds developing directly into sporophytes and the plants do not show any dependence on water.

Although the sporophyte is long-lived, the fern gametophyte is usually ephemeral. It develops in a microenvironment characterized by little competition from other plants (including even mosses and algae); exposed humus, decomposing plant materials, or fresh mineral surfaces; deep to moderate shade; and a humid atmosphere. Even ferns whose sporophytes tolerate sun and drought tend to have these requirements for their gametophytes to grow well. On rocks, for example, the gametophytes form in protected crevices in which light is minimal and moisture maximal. Because of their requirements for exposed soil, development of fern gametophytes is promoted by damage to mature vegetation, by conditions such as fallen trees in the forest, flooding, and deep erosion. Prothallia are observed in nature most commonly upon shaded soil banks, along streams and upon rotting logs.

The bulk of reproduction of ferns is vegetative, taking place in the sporophytic stage by rhizomes and in some cases by root or leaf proliferations. In fact, reproduction by vegetative means probably accounts for the bulk of fern plants in the world. The sexual cycle, including spores and independent gametophytes, is probably important primarily in invading new habitats, extending the plant's geographic range, and creating genetic variation.

Sporophyte

A fern sporophyte consists of a stem, which is often called a rhizome, even if it is above ground or forms a 20 m tall trunk, as in some tree ferns.

From the rhizomes, leaves or fronds grow, which in ferns are prominent - megaphylls (in contrast to lycopods, which bear microphylls: simple leaves with single unbranched veins). Leaves when emerge are circinately coiled. The petiole is used as a diagnostic character and in particular the number of vascular strands is important. Leaves may be simple or highly divided into pinna and pinnules.

Fertile leaves or sporophylls bear spore-producing structures, the sporangia, which are organized in groups, the sori. Usually sori are on the lower side of the sporophyll, and in some groups, specialized structures such as indusium (and paraphyses) remain associated with the sporangia and sori.

Gametophyte

When the spore wall breaks the spore germinates to form a gametophyte. Emerging first from the spore is a nongreen rhizoid (root-like organ), which attaches the plant to the growing surface, and a green single cell—the mother cell that gives rise to the rest of the gametophyte (Fig. 5.2A-D). At first, in most homosporous ferns, growth is in the form of a single filament, and it may continue in this fashion if lighting conditions are weak. Under optimum conditions of light, the gametophyte becomes a two-dimensional sheet of cells and later an apical cell is organized. The prothallus has an apical notch surrounded on either side by the prothallial wings—flat plate-like protrusions, one cell thick. The average size of the gametophyte at the time of fertilization is approximately 2 to 8 mm (0.08 to 0.32 inch) long and up to 8 mm wide.

The gametophytes of ferns showing heterospory are endosporous; that is, they do not emerge in germination and fail to grow beyond the confines of the spore walls. The microspores produce antherozoids in antheridia, and the megaspores produce eggs in archegonia. The endosporous gametophytes are much reduced. The male gametophyte in the microspore is made up of the equivalent of one antheridium and its complement of antherozoid. The female gametophyte, although considerably larger in size, is equally reduced in a morphological sense. The megaspore remains filled by stored nutritive materials and tissues formed around the base of the archegonia.

The vegetative phase of the gametophyte in these forms has been practically eliminated, and the developing embryo in the megaspore lives on stored food materials, photosynthetic efficiency of spore being poor. The differentiation between male and female gametophytes ensures cross-fertilization.

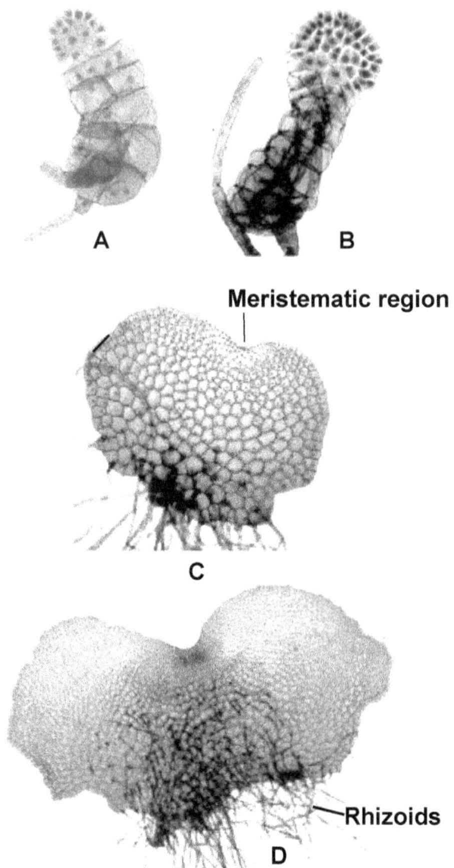

Figure 5.2A-D: Spore germinates to form a small filamentous structure (A) and rhizoids get initiated (B). Later a meristematic zone gets differentiated (C) which after growth results in a heart-shaped prothallus (D).

Specialized forms

From a basic type of gametophyte, somewhat a heart shaped, a number of highly specialized forms have evolved that are characteristic of certain genera. Ribbon-like and strap-shaped gametophytes are known especially among tropical rainforest ferns such as *Vittaria*, *Grammitis*, and *Hymenophyllum*. These are usually irregularly and extensively branched, forming large masses of intertwining ribbons. Filamentous (thread-like) gametophytes are known in the genera *Trichomanes* (*Hymenophyllaceae*) and *Schizaea* (*Schizaeaceae*). Tuber-like gametophytes occur in several groups—

e.g., the families Ophioglossaceae (all members), Schizaeaceae (*Actinostachys*), and Gleicheniaceae (*Stromatopteris*). All are non-green underground plants that have close associations with fungi, being mycoheterotrophic (i.e., dependent on fungi for nutrition from organic material in the soil). They commonly occur 5 to 10 cm (2 to 4 inches) deep in the ground.

Vegetative reproduction

Vegetative propagation of some photosynthetic fern gametophytes is accomplished by continued growth and fragmentation, but this does not spread the gametophyte very far. Some ferns (*Vittaria, Grammitis,* and the family Hymenophyllaceae) produce specialized filaments, or gemmae, that break off and are carried away by some agency such as water droplets, wind, or possibly insects or spiders to initiate new colonies. This mode spreads the plant to certain distances.

Asexual reproduction

In the year 1874, Farlow discovered the natural occurrence of apogamy in *Pteris cretica, Dryopteris, Osmunda* and *Adiantum* and later in 1878, Heinrich Anton de Bary, termed this type of asexual reproduction in ferns as Apogamy**.** In addition to the normal life cycle using meiosis to generate spores and sexual union to form zygotes, in nature, some fern species can switch from one generation to another by different pathways termed apospory and apogamy. In the first case, a gametophyte is generated from sporophytic cells without meiosis, and in the second one, a sporophyte is generated from gametophytic cells without fertilization. In the laboratory it is possible to induce both pathways using specific culture conditions. The independence of the two generations in ferns and the ease of switching from one generation to the other through these pathways offers a system suitable for studying how each generation is initiated. This developmental plasticity of crossing generation barriers, i.e., meiosis and fertilization, is not unique to ferns and is manifested in the complex pathways leading to apomixis in some seed plants. The asexual pathways in ferns are compared and contrasted with the current understanding of apomixis in seed plants.

Sexual reproduction

The antheridia may be sunken (as in the families Ophioglossaceae and Marattiaceae) or they may be protruding. They vary in size from antheridia with hundreds to those with only 12 or so antherozoids. The egg-producing organ, the archegonium has a neck that consists of four rows of cells containing the neck canal cells, the venter canal cell and the egg.

Fertilization is as in other groups of archegoniate, attained by the ejection from the antheridia of antherozoids which swim through free water toward chemotrophic substances secreted by the archegonium. The antherozoids swim in the neck canal and penetrate the egg. When the egg is fertilized, the embryo develops within the expanding venter.

Embryo

The zygote undergoes characteristic cell divisions to form the embryo, which remains encapsulated in the gametophyte until it emerges out and becomes an independent plant. In a few species the first division of the zygote is transverse. The inner cell grows producing the stem and first leaf, and the outer cell divides to form a foot, a mass of tissue that exists as part of the embryo and disappears when its function, presumably absorption, is completed. The root appears later within the stem and grows outward. In majority of all other known ferns, the zygote divides neatly into four quadrants, the first division approximately parallel to the long axis of the archegonium and the following division at right angles. This results in initial cells that give rise to four organs: the outer cell (i.e., toward the growing apex of the gametophyte and the neck of the archegonium) becomes the first leaf, the inner cell the stem apex, the outer back cell, the first root, and the inner back cell, the foot.

The young sporophytes of ferns remain attached to the gametophytes for varying lengths of time, absorbing nutrients from the gametophyte through the foot. Once the sporophyte develops into independent individual and the root penetrates the soil, the gametophyte shrivels and is no longer required.

Stem

Stem growth is initiated by one to several large apical cells. Leaves and leaf bases formed of old sclerified cells play a major role in the protection of fern stems. The old leaf bases may also serve as food-storage organs. In most species the stems are indeterminate in growth and thus can theoretically continue to grow indefinitely.

Most fern stems also are covered with trichomes, or scales which are so distinctive that they serve identification and classification of species. There may be other epidermal outgrowths such as simple glands, simple unbranched non-glandular trichomes, dendroid trichomes (branching filaments), and scales (flat cell plates) of many patterns. Scales also known as paleae, have a cell plate, two or more cell rows wide, at least at the base. The hairs on the other hand, each generally consist of a single row of cells. Transitional stages are also seen in some species.

The cortex is composed mostly of storage parenchyma cells (a relatively generalized cell type). There is a strong tendency for the outermost cortical cells to become darkly pigmented and thick-walled especially in the rhizome.

The steles exhibit somewhat diverse patterns (Fig. 5.3A-I). Most common ferns possess a dictyostele, consisting of vascular strands interconnected in such a manner that, in any given cross section of stem, several distinct bundles are observed. These are separated by regions filled with parenchyma cells known as leaf gaps. There are, however, numerous siphonostelic ferns, in which the gaps do not overlap and a given section shows only one gap. Some ferns possess protostele, while complex stelar patterns such as polycyclic dictyostele, in which one stele occurs within another stele, are known in some species (*Pteridium*). Large strands of fibre-like cells running between the two steles form mechanically specialized hard tissue, or sclerenchyma.

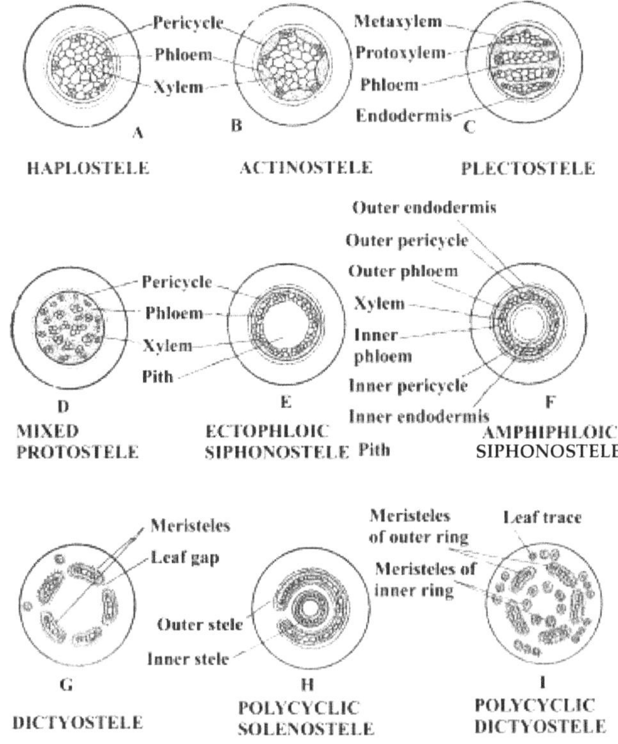

Figure 5.3 A-I: Stele types in stem.

Root

Fern roots are generally thin and wiry, although some are fleshy and either slender (in the Ophioglossaceae) or thick as much as 13 mm (0.5 inch) in diameter (e.g., *Acrostichum* and *Marattia*). Certain tropical ferns have elaborately hairy roots whose surfaces are covered with locks of silky golden or brown root hairs. In certain tree ferns (e.g., *Cyathea* and *Cibotium*) and in the royal ferns (*Osmunda*), the entire stem surface is covered with roots masses. The roots have been put to use in nurseries. The large dead tangles of tree fern roots can be cut into various shapes suitable for attaching epiphytic greenhouse plants. Pieces of such root masses have also been useful in horticulture for cultivating orchids and bromeliads.

Leaf

The leaf plan in practically all ferns is pinnate. From this basic type a broad diversity of forms evolved (Fig. 5.4A-H). Some ferns have palmate leaves (with veins or leaflets radiating from one point), and some, such as the staghorn ferns, have secondarily evolved falsely dichotomous leaves. In some genera (e.g., *Lygodium* and *Salpichlaena*) the main leaf axis (rachis) twines about on shrubs and small trees, sometimes reaching 20 metres (65 feet) in length. The fern leaves can usually be distinguished readily from the scale-like or awl-like leaves (microphylls) of club mosses. A few ferns (e.g., sword ferns, *Nephrolepis*; *Jamesonia*; *Salpichlaena*; and climbing ferns, *Lygodium*) have members with more or less indeterminate (i.e., continuous) leaf growth accomplished by periodically quiescent buds. In most ferns leaves grow from apical cells, and these delicate embryonic cells are protected by the curled-over spiral of the crosier (unrolling leaf tip) and by trichomes or scales. When the blade formation is complete, the embryonic tip becomes inactive. In overall, length of mature leaves varies from 1 or 2 mm (0.04 or 0.08 inch) in certain filmy ferns (Hymenophyllaceae) to 30 or more metres (100 feet; family Gleicheniaceae). In terms of overall size, the most massive leaf is that of the elephant fern (*Angiopteris*), with length more than 5 metres (16 feet) and petioles 15 cm (6 inches) in diameter.

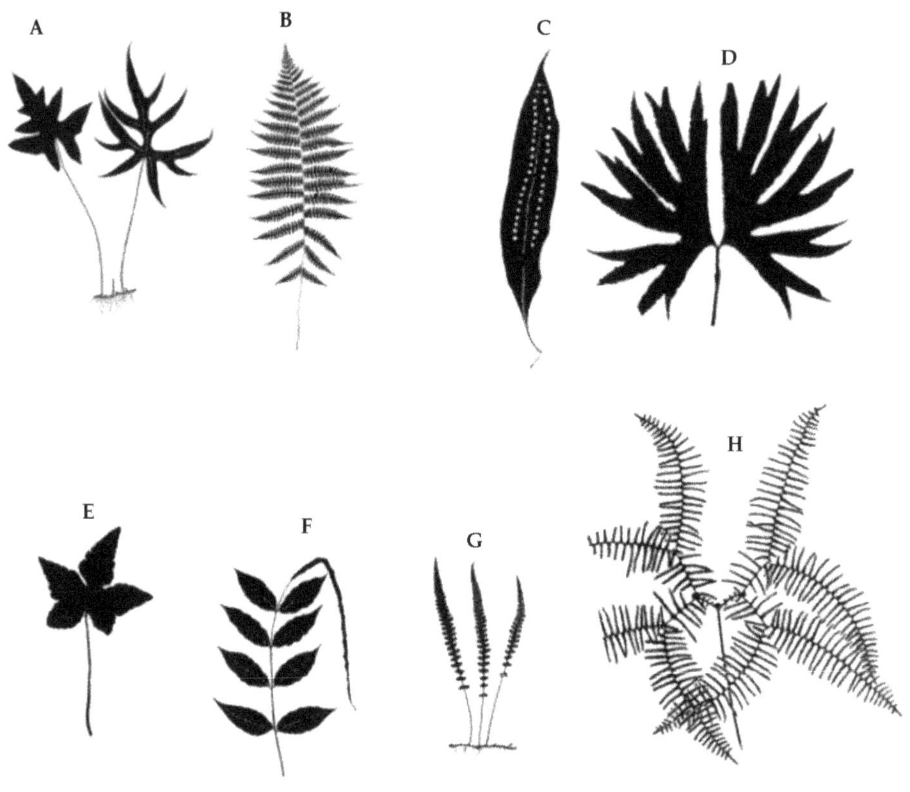

Figure 5.4A-H: Leaf types in ferns.

(A) *Dryopteris nobilis*, pedate laminae. (B) *Deparia acrostichoides*, lamina 1-pinnate-pinnatifid. (C) *Microgramma megalophylla*, simple and entire lamina. (D) *Dipteris conjugata*, lamina divided into two halves from top of petiole. (E) *Hemionitis palmata*, palmate lamina. (F) *Bolbitis heteroclita*, 1-pinnate lamina with elongate apical segment proliferous at tip. (G) *Diplazium tomitaroanum*, pinnatifid leaf. (H) *Gleichenia microphylla*, pair of opposite pinnae.

Whether a given leaf is pinnate (compound) or is undivided (simple e.g., *Asplenium scolopendrium* var. *americanum*) is of considerable value in identification of similar fern species. The pinnae themselves may also be lobed or truly divided with stalked segments; and the resulting segments, the pinnules, may also be lobed or divided. Depending on the degree of cutting, leaves are described as simple, once divided, twice divided, thrice divided, and so on.

Generally, the patterns of the leaves are pinnate, and the veins are free; that is, they all diverge and never coalesce, either along their sides or at the ends. Nevertheless, there are numerous fern groups in which netted, or reticulate, venation is found. These have vein patterns like those of other ferns for the most part, except that various systems of networks and areolae (areas enclosed within loops of veins) have developed between the major, pinnately arranged veins. One of the more striking reticulate pattern is that in which each loop or areola contains one or more free included veinlet, as seen in various members of the family Polypodiaceae. Another is the herringbone pattern, believed to result from an evolutionary concrescence (growing together) of pinnae, as shown by certain tree ferns (*Cyathea*), lady ferns (*Athyrium*), and marsh ferns (*Thelypteris*).

Fern leaves vary in the relationship of the petiole to the blade. Many strap-shaped leaves essentially have no petiole and are described as sessile; broad, ovate, or triangular leaves commonly have a pronounced leaf stalk, called a stipe, and are termed petiolate or stipitate. Narrowly elongated leaves in ferns are usually erect, spreading, or, in certain epiphytes, pendent. Leaves that are broadly ovate or triangular tend to be oriented at right angles to the incident light.

Anatomically, the petioles show nearly as much diversity in cross-sectional pattern as do the stems. The simplest vascular strands of fern petioles are commonly crescent-shaped single bundles. In more elaborate petiolar patterns, the crescent takes on the form of the Greek letter omega (Ω), opening adaxially (i.e., upward or toward the central axis of the plant). Double-stranded ferns (the omega now divided into two parts and unconnected below) are usually associated with more specialized genera (e.g., *Athyrium* and *Thelypteris*). Any of the generalized patterns may exist as broken-up strands where the small leaves have only three to nine strands, while the large leaves of tree ferns have many. Petiolar vascular bundle shapes have been of certain value in separating fern genera and families.

At the tissue level there is an upper and a lower epidermis, the latter with many stomata. Epidermis is followed by mesophyll, usually with large intercellular spaces. In thicker leaves, the mesophyll is composed of palisade cells—elongated cells arranged parallel and oriented with the long axis vertical to the leaf surface. The veins range from the massive major midribs, which have well-defined xylem, phloem, pericycle, and endodermis, to the delicate capillaries represented by a few tracheids that sometimes end in a cluster of somewhat modified tracheids. Underneath a sporangium or sorus, the veins may become dilated and multilayered (so-called fertile veins). In many ferns nearly all of the photosynthesis is accomplished by the epidermis, the mesophyll having been eliminated in evolution. An example is the common maidenhair fern (*Adiantum pedatum*), the blade of which, between veins, is mainly made up of only two layers, the upper and the lower epidermis, in which most photosynthesis occurs.

Numerous ferns have at least a few microscopic trichomes, which are usually glandular and appressed to the blade surface.

The fern leaf vernation is circinate; that is, the leaf unrolls from the tip rather than expanding from a folded condition. It also differs in its venation, which usually is free or simply reticulate made up of areolae containing numerous branched, free-ending veinlets.

SPORANGIUM AND SORUS

The sporangium

The spore-producing structures, in ferns range from globose sessile (non-stalked) organs more than 1 mm (0.04 inch) in diameter to microscopic stalked structures, the capsules of which are only 0.3 mm (0.01 inch) in diameter. The former known as eusporangia arise from several cells, while the leptosporangia, the latter type arise from a single cell. Eusporangia occur in the classes Psilotopsida and Marattiopsida, and leptosporangia occur in the majority of the species in the class Polypodiopsida. There are, however, many forms intermediate between the two types of sporangia, and these are known in various primitive species of the Polypodiopsida, such as members of the family Osmundaceae (Classes mentioned as under Smith et al., 2006).

The capsule wall in eusporangia tends to be relatively massive, made up of two layers or more. In leptosporangia, on the other hand, the wall is thin composed of one layer of cells. Opening of the capsule in eusporangia, such as those of the genus *Botrychium*, is accomplished by separation along a well-differentiated line of dehiscence (opening); but in most typical leptosporangia, a few stomial ("mouth") cells differentiate and the process of dehiscence tears the cells apart more or less irregularly.

The opening process in eusporangia is the result of a generalized stress on drying walls with differentially thickened cells. There is no mechanism to disperse the spores, and they are simply carried away by the wind. In contrast, leptosporangia display more or less specialized annuli, usually consisting of a single row of differentially thickened cells. Apparently, the mechanical force for opening and spore dispersal derives entirely from these annular cells. The stress imposed by the drying of the annular cells results in the collapse of the outer walls of the cells, thus straightening out the annulus and ripping the soft lateral cells of the capsule apart. As the annular cells continue to collapse by the cohesive forces of the increasingly tense water molecules within, the spore case completely opens. Finally, the water film between the outer walls of the annular cells breaks, and the entire annulus snaps back to its original position, tossing the spores into the air. This is the catapult mechanism (Fig. 5.5A-C).

Most primitive sporangial types are stalkless, or sessile with merely a slightly raised multicellular area at the base of the sporangial capsule in some cases. In typical leptosporangia, however, there commonly are well-developed long and narrow stalks (e.g., as in *Davallia* and *Loxoscaphe*), made up of only one or two rows of cells and often 1.7 to 2 mm (0.07 to 0.08 inch) in length.

The sporangial evolution is from solitary large capsules to increasingly elaborate groupings of smaller sporangia known as sori. These changes are accompanied by the appearance of structures such as paraphyses and indusia. While paraphyses are sterile structures that grow among or on the sporangia, indusia are papery tent-like structures that cover the sori (clusters of sporangia).

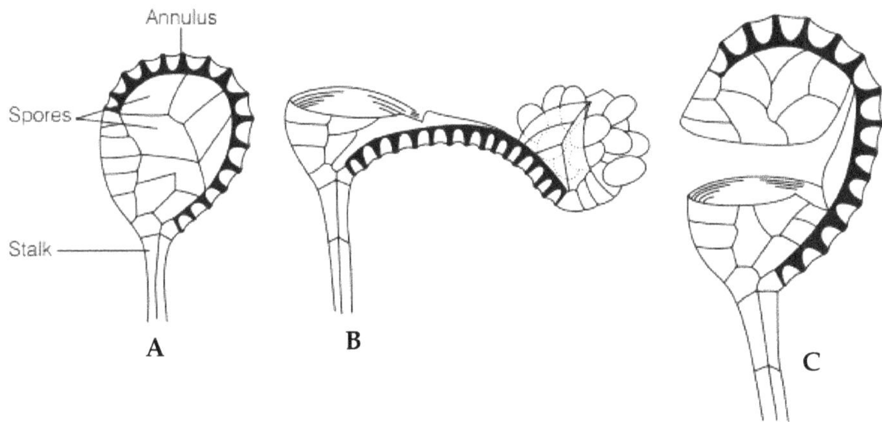

Figure 5.5A-C: Catapult mechanism of spore release.

The sorus

Besides reduction in the size of single sporangium, more complex aggregations of sporangia known as sori are seen to occur in certain species. They produce sporangia of all ages, older ones being pushed aside as new ones reach maturity. In some taxa, most notably belonging to Marattiales, eusporangia are fused together, forming a complex synangium. Most remarkable are perhaps the heterosporous ferns in which a single sporophyte produces a smaller number of large spores, megaspores, that develop into female gametophytes and, more typically, numerous microspores that develop into male gametophytes. In ferns, this condition is found only in two groups of aquatics: Marsileaceae and Salviniaceae. It is clearly an adaptation to their aquatic environments, where the female gametophyte develops inside the megaspore and is thus protected from the surrounding water. These heterosporous ferns have often been placed among 'fern allies' along with the unrelated heterosporous lycopods (Selaginellaceae, *Isoe¨taceae*). However, heterospory has evolved independently in these lineages and is not necessarily associated with aquatic habitats. Spores in heterosporous ferns are formed in special structures called sporocarps.

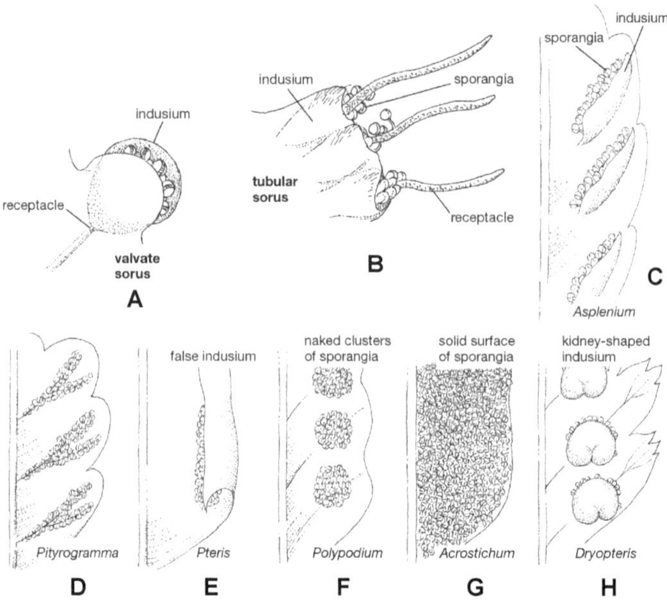

Figure 5.6 A-H: Fern Sporangia.

In ferns, spores are contained within cases called sporangia that are located on the underside of leaves.

The stages in progressive evolution of sori can be depicted as follows: (1) simple clusters of sporangia, these are more or less coalesced (family Marattiaceae) or separate (Gleicheniaceae), all of them maturing at the same time, (2) gradate clusters of sporangia, the outermost ones maturing first, the innermost last, and (3) mixed clusters of sporangia, all ages present, the younger ones arising from the same meristematic zones as the older ones. The adaptive significance of this change is probably related to the duration of spore production, the mixed character of the more advanced sori extending the period beyond that of solitary sporangia or of simple, simultaneously maturing sori.

Sporangia, especially the sori have traditionally provided the most important characters for fern classification. Occasionally they cover the entire lower lamina of the blade, especially towards its apex, and the condition is called an acrostichoid sorus (Fig. 5.6 A-H). Where present, ferns have one of the following six soral arrangements: (1) A linear arrangement of sporangia along veins, avoiding the leaf area between the veins, is found in many fern genera, especially in the genus *Pityrogramma* (2) A line of sporangia along the leaf edge, protected usually by a rolled-over and modified laminar margin, is represented by *Pteris* (3) Round and naked sori (i.e., without an indusium) are found in *Polypodium* (4) An arrangement of large sori that usually expand over the entire undersurface of the blade or pinna is represented by *Acrostichum*. Such sori probably arose by the fusion of smaller clusters of sori (5) A linear or oblong sorus along a vein covered from one side by a narrow indusium, which is represented by *Asplenium*, and (6) A sorus that is round but covered with a kidney-shaped or shield-like indusium, which is represented by *Dryopteris* (Fig. 5.6).

The indusium

Protection of the sporangial cluster from exposure, drying, and other natural hazards is accomplished by the formation of the sori in grooves or pockets or by the production of various forms of covers. One such protective cover is the so-called false indusium, a rolled-over leaf margin under which sporangia form and mature. The true indusium constitutes a more or less papery covering over the sorus. A widespread type of indusium among members of the family Cyatheaceae is shaped like a cup, which arises around the base of the sorus, often enclosing the sorus until the sporangia are mature (e.g., *Cyathea*). In some genera, marginal sori are protected by a two-lipped, or valvate, indusium (e.g., *Dennstaedtia,Dicksonia*, and *Hymenophyllum*). When sori fuse laterally to form continuous lines, or coenosori, indusia also tend to fuse.

Paraphyses

Approximately one-third of fern species have paraphyses, sterile hairs or scales intermixed with the sporangia with a protective function. Paraphyses may occur on the sporangial stalk or capsule. Paraphyses structure has proved to be helpful in taxonomic data.

Spores

Most ferns are homosporous, each plant having spores of one shape and size, usually 30 to 50 micrometres in length or diameter, although some reach more than 100 micrometres. A few fern families which are heterosporous, however, have dimorphic spores, small ones (microspores) and large ones (megaspores).

Ferns display a wide diversity of spore types in terms of shape, wall structure, and sexuality, which have great value in determining taxonomic relationships. The basic spore shape among ferns is tetrahedral; the proximal face (the one facing inward during the tetrad, or four-cell, stage following reduction division, or meiosis) is made up of three sloping planes, and the distal, or outer, face consists of a single rounded surface. The tetrahedral structure is commonly obscured in so-called globose spores, the walls of which are thin and soft. Often, the wall is composed of exospore (outer spore layer) only, there being no additional jacket, or perispore. The wall may be either unsculptured, smooth or provided with a variety of sculptured patterns. The tetrahedral spore is formed by cellular divisions in two different planes at right angles to one another during meiosis in the spore mother cell.

In contrast, the bilateral spore type of many fern species is formed by successive cell divisions in planes along a single axis during meiosis of the spore mother cell. This results in a linear (monolete) scar running parallel to the long axis of the spore. Most bilateral spores in ferns are bean-shaped and jacketed by a perisporial layer, a distinctive covering of the outer wall.

Cytogenetics

The chromosomes of ferns tend to have high base, or *x*, numbers, ranging from approximately 20 to 70, with the majority between 25 and 45. The familiar genus *Osmunda*, for example, has $x = 22$, *Pteris* 29, *Asplenium* 36, *Dryopteris* 41, *Botrychium* 45, and *Pteridium* 52. *Ophioglossum reticulatum* has 1,440 chromosomes, the highest number in any organism known to science. Among homosporous ferns, filmy fern (*Hymenophyllum peltatum*), lowest number $x = 11$ is reported. Among heterosporous ferns, however, the situation is conspicuously different, and all have low base numbers (*Marsilea*, $x = 10$, 13, or 19; *Salvinia*, $x = 9$; *Azolla*, $x = 22$).

However, genetic studies have shown that despite their high chromosomal base numbers, most species act functionally as diploids rather than as polyploids. Though ferns still have relatively high levels of polyploidy, but these polyploids are all of relatively recent origin with approximately 45 percent of the extant species of ferns being neopolyploids.

The base chromosome numbers (indicated by the symbol *x*) have been used for classification purposes.

Simple polyploid series with multiples of the base number are prevalent among ferns, with a few species reported to have forms or races that are diploid (with two times the base number of chromosomes), tetraploid (four times), or hexaploid (six times). For example, the fragile fern complex centred on *Cystopteris fragilis* has species with the diploid number represented by $2n$—equal to two, four, and six times the base number of $x = 42$; or $2n = 84$, 168, and 252. Species with both diploid and tetraploid forms are common among widespread ferns.

Hybridization–In certain fern genera, such as spleenworts (*Asplenium*), wood ferns (*Dryopteris*), and holly ferns (*Polystichum*), hybridization between species (interspecific crossing) may be so frequent as to cause serious taxonomic problems. The majority of hybrids are sterile and reproduce, if at all, only by vegetative propagation.

In some genera of ferns, there is a natural trend toward the production of unreduced spores in a small minority of the sporangia. When these spores with twice the starting number of chromosomes germinate, the resulting gametophytes have the same number of chromosomes as the maternal sporophyte. Often gametes (2n) from the rare diploid gametophytes cross-fertilize with reduced gametes from natural spores (n) produced by meiosis. The sporophyte that results from the union of diploid and haploid gametes in such cases is triploid and always adopts an apomictic lifestyle. In sterile fern hybrids apogamy, leads to the formation of unreduced spores (with the $2n$ number of chromosomes) that germinate into normal-appearing male gametophytes. The hybrid gametophytes simply bud off a new sporophyte clonally, and there is a direct conversion from gametophyte to sporophyte generation. Apogamous ferns are known in a number of genera of higher ferns, including *Adiantum*, *Asplenium*, *Cheilanthes*, *Dryopteris*, *Pellaea*, *Polypodium*, and *Pteris*.

Besides apogamous hybrids, there are sexual "allopolyploid hybrids," which are believed to have originated by doubling of the chromosomes of sterile crosses (but without the shift to an apogamous lifestyle). Both apogamous and allopolyploid hybrids may enjoy wide geographic ranges occurring in as great abundance as normal species. Both types are also capable of creating additional hybrids by backcrossing (to the parent species) or by crossing with other species.

Curiously, in spite of the high number of ferns that are epiphytic, nearly all the fern hybrids are terrestrial or epipetric (growing on rocks); hybridization being rare among epiphytes.

Fossil record

The Carboniferous Period (358.9 million to 298.9 million years ago) was a time of great evolutionary processes in ferns, and diversification of their form was phenomenal. The ferns were at the pinnacle during the Carboniferous Period (the age of ferns) as they formed the dominant part of the vegetation at that time. Though most of those groups are now extinct, but some of them later on evolved into the modern ferns.

Modern ferns, however, are relatively uniform in basic structure, sharing a large number of characteristics. All the living families, with the exception of the primitive classes–Psilotopsida, Equisetopsida, and Marattiopsida—possess a ground plan of correlated characteristics that seems clearly to bind them together as an assemblage that is monophyletic (i.e., having one evolutionary line). The construction plan of modern ferns is so distinctive that the vast majority of them can be recognized immediately as members of this group. However, one or two extant genera can be traced to direct ancestors in the Carboniferous. There are however, no immediate ancestors of modern ferns in the fossil record.

The earliest true ferns arose during Carboniferous times, or perhaps a few in Devonian, and have been classified in five families—Marattiaceae, Equisetaceae, Osmundaceae, Gleicheniaceae, and Schizaeaceae. Several extinct groups of the Carboniferous Period and the Permian Period (298.9 million to 251.9 million years ago) that followed such as Coenopteridaceae, Anachoropteridaceae, Tedeliaceae, Sermayaceae, and Tempskyaceae form the related lines of evolution, but the intermediate examples are missing.

By the Triassic Period (beginning 251.9 million years ago), some of the modern fern families were well established. There are fossil records of the families Osmundaceae, Equisetaceae, Marattiaceae, Schizaeaceae, Matoniaceae, Dipteridaceae, Cyatheaceae, Marsileaceae, and Salviniaceae. However, the bulk of the modern fern species did not diversify until the Cretaceous Period (145.0 million to 66.0 million years ago), at or slightly after the time of the great diversification of angiosperms.

EVOLUTIONARY DEVELOPMENT

Despite a relatively large number of theories, the actual origin of the vegetative organs of ferns is still unknown.

The leaf is equally or even more problematic as to its ultimate origin. Various hypotheses have been offered, of which the telome theory (that the leaf arose from fusions and rearrangements of branching stem systems) and the enation theory (that the leaf arose from simple enations, or outgrowths) are the two most popular. Leaves of most modern ferns, bear characteristic fiddleheads, acropetal growth (i.e., "seeking the apex," the leaf tissues maturing from the base toward the tip, where the youngest tissues are produced), and pinnate structure, which are quite distinctive in the vascular plants. They differ in numerous respects from those of conifers, and of flowering plants.

CLASSIFICATION

The classification of ferns has long proved problematic and has always been fluid. Some past works had grouped them with other non-seed bearing and spore producing vascular plants such as lycopsids (club mosses, spike mosses and quillworts), sphenophytes (horsetails) and psilophytes (whisk ferns). Such works and classification schemes referred to them as "ferns and fern allies" respectively. However, the recent classifications place lycopsids in a separate group while ferns are now firmly placed within the Monilophytes. This monophyletic group and its sister group, the spermatophytes (seed plants and progymnosperms) form the two extant (not-extinct) clades of the euphyllophytes. Euphyllophytes and lycophytes probably diverged in the late Early Devonian, while the monilophyte and spermatophyte clades probably separated during the Middle Devonian.

Tracing Classification with Early Attempts

Linnaeus (1753) recognized in ferns 15genera followed by Swartz (1806) who recognized 38 genera, more than twice as many as were recognized by Linnaeus. About 20 years later, the number had again more than doubled, to 79 (Desvaux, 1827) and by the middle of the 19th century, when new plants were being discovered and described, the number of recognized fern genera rose to 134 (Hooker, 1842), 143 (Smith, 1842a, b, c, d, 1843), 176 (Presl, 1836, 1843, 1845, 1847, 1851), and then 181 (Fée, 1852). It was the classification of Hooker (1846–1864), which recognized just 68 morphologically heterogeneous genera, discounting the works of his predecessors. As time passed, new approaches, new data, and new insights again led to global acceptance of greater numbers of genera, with 216 recognized by Christensen (1938), 251 by Ching (1940), 308 by Copeland (1947), 404 by Crabbe et al. (1975), and 435 by Pichi Sermolli (1977). The compilation edited by Kramer & Green (1990), published just prior to the molecular phylogenetic revolution listed, 217 genera. However, a growing consensus on fern systematics has been developing during the last decade as a result of the cladistic analyses of morphology (including fossil taxa), comparisons of antherozid ultrastructure, and DNA sequencing; the DNA analyses have proven particularly compelling. A global consortium of nearly 100 systematists has published a community-derived classification for extant pteridophytes (PPG I, 2016) and recognized 18 lycophyte and 319 fern genera in 51 families, 14 orders, and 2 classes. The work is based on the synthesis of data from morphology and molecular studies.

Modern Ferns

The most widely accepted previous schemes placed three groups of extant ferns (Marattiales, Ophioglossales and leptosporangiate ferns) within the "true ferns" while two other groups, sphenophytes (horsetails) and Psilotales (whisk ferns) were regarded as more distantly related "fern allies" that may or may not be "true ferns". The designation "true fern" is now no longer valid because of some Ophioglossales, and one "fern ally", Psilotales, form one lineage of Monilophytes while leptosporangiates and Marattiales ("true ferns"), and sphenopsids, form the other.

Critical appraisal

The classification of ferns has been in a state of flux over the past several decades. However, advances in molecular techniques have resulted in first phylogenetically based system of classification at the family, order, and class levels. In 2016, the Pteridophyte Phylogeny Group published their first scientific consensus of fern and lycophyte phylogeny, known as PPG I. Though it has received huge acceptance but within some families, the circumscription of genera remains controversial.

The class Polypodiopsida of PPG1 contains four subclasses of vascular plants that reproduce by spores. In addition to Polypodiidae (the true ferns as traditionally understood), the class includes the subclasses–Equisetidae, Marattiidae, and Ophioglossidae, which in some earlier classifications were considered "fern allies" (an expression that has fallen out of favour) rather than true ferns. The class Lycopodiopsida, or lycophytes, consisting of three families, Lycopodiaceae (club mosses), Isoetaceae (quillworts), and Selaginellaceae (spike mosses), formerly included in the fern allies, are now considered the most primitive vascular plants and only distantly related to the ferns.

A brief account of PPG 1 classification with salient features of families: Numbers given for the species are only rough approximations of living groups.

CLASS POLYPODIOPSIDA (FERNS)

SUBCLASS OPHIOGLOSSIDAE

ORDER OPHIOGLOSSALES

FAMILY OPHIOGLOSSACEAE (adder's tongue, grape ferns, moonworts)

Plants usually with somewhat fleshy stems and roots; leaves divided into sterile and fertile segments, these variously entire to highly divided, not developing through circinate vernation, the base more or less clasping the stem; eusporangiate (with unstalked, globose sporangia); gametophytes subterranean, not green; 4 or more genera (*Botrychium*, *Helminthostachys*, *Ophioglossum*, and *Mankyua*) with about 80 species.

ORDER PSILOTALES (Whisk ferns)

FAMILY PSILOTACEAE

Plants lacking true roots, stems dichotomously branched, protostelic (the stele lacking pith and leaf gaps); leaves reduced to minute scale-like outgrowths (enations) without veins, or small, flattened, and undivided, with a single midvein (microphylls), not developing through circinate vernation; sporangia eusporangiate in fused clusters (synangia) of 2 or 3, situated along the aerial stems; gametophytes similar in appearance to the rhizomes of the sporophytes, subterranean, not green; 2 genera (*Psilotum*, *Tmesipteris*) with 12 species found in tropical and warm temperate regions nearly worldwide.

SUBCLASS EQUISETIDAE

ORDER EQUISETALES (Horsetails)

FAMILY EQUISETACEAE

Stems with whorled branches, longitudinally ridged, hollow between the nodes, with characteristic rings of longitudinal canals; leaves whorled, reduced to a ring of small scale-like structures, these with a single vein, often papery and not green; sporangia eusporangiate, positioned under small umbrella-shaped branchlets that are grouped into complex terminal, cone-like structures (strobili); spores green, with 4 flattened thread-like appendages (elaters) that aid in dispersal; 1 modern genus (*Equisetum*) with 15 species, distributed nearly worldwide.

SUBCLASS MARATTIIDAE

ORDER MARATTIALES (Giant ferns)

FAMILY MARATTIACEAE

Leaves pinnately divided, pulvinate (enlarged or swollen at attachment point of leaflets) in living genera, and with well-developed, fleshy stipules (appendages at leaf base); sporangia eusporangiate, in *sori*, or more or less coalescent in synangia (clusters); homosporous; mostly massive, fleshy ferns; 6 modern genera (*Angiopteris*, *Christensenia*, *Danaea*, *Eupodium*, *Marattia*, and *Ptisana*) with about 150 species, widely distributed in tropical regions.

SUBCLASS POLYPODIIDAE (Leptosporangiate ferns)

ORDER OSMUNDALES

FAMILY OSMUNDACEAE (Royal ferns)

Plants in soil, often in wetlands; rhizomes often stout, usually erect or ascending, occasionally trunk-like; leaves 1 to 4 times pinnately divided; the petiole thickened at the base; sporangia often on contracted, sometimes nongreen segments of entirely separate fertile leaves or fertile portions of otherwise vegetative leaves, less commonly along veins of unmodified leaf segments, maturing nearly simultaneously, intermediate in spore number between eusporangia and leptosporangia, the annulus a lateral patch of thick-walled cells; 6 genera (*Claytosmunda*, *Leptopteris*, *Osmunda*, *Osmundopteris*, *Plenasium*, and *Todea*) and 20 modern species, distributed nearly worldwide.

ORDER HYMENOPHYLLALES

FAMILY HYMENOPHYLLACEAE (Filmy ferns)

Mostly rainforest epiphytes; tiny ferns with blades only 1 cell thick between veins; spores globose, green; gametophyte ribbon-shaped or filamentous, gemmiferous; principal genera are *Hymenophyllum* and *Trichomanes*; 9 genera (different authorities give anywhere from 2 to more than 28) with some 600 species found in tropical regions around the world, a few species extending into temperate areas.

ORDER GLEICHENIALES

FAMILY GLEICHENIACEAE (Forking ferns)

Plants in soil or on rocks; rhizomes creeping; leaves mostly sprawling over other vegetation, falsely dichotomous, the segments mostly narrowly lobed; sporangia with oblique annuli and clustered in simple sori lacking indusia; stems creeping, protostelic (its stele lacking pith and leaf gaps); *Gleichenia, Dicranopteris*, and 4 other genera with about 125 species, distributed in the tropics.

FAMILY DIPTERIDACEAE (Umbrella ferns)

Plants in soil; rhizomes long-creeping, hairy; leaf blades usually palmately divided into 2 or more lobes, the veins of at least the vegetative fronds forming a dense network; sori scattered on or covering the leaf undersurface, each associated with 2 or more veins, thick, stalked, the annulus vertical or slightly oblique; 2 genera, *Dipteris*, with about 11 species distributed mostly in the Paleotropics, and *Cheiropleuria*, with 1 species (*Cheiropleuria bicuspis*) in the Paleotropics.

FAMILY MATONIACEAE

Plants in soil; leaves either fan-like, with lobed, narrow segments, or climbing, with long midribs; sporangia with oblique annuli, the simple sori covered by a thick umbrella-shaped indusium-like structure; 2 genera (*Matonia* and *Phanerosorus*) with 4 species, distributed in the Paleotropics.

ORDER SCHIZAEALES

FAMILY SCHIZAEACEAE

Leaves more or less grass-like, with a long petiole and a linear or fan-shaped blade; veins dichotomously branching; sporangia dense on specialized slender lobes of the ultimate segments; the annulus a subapical ring of thickened cells; 2 genera (*Schizaea* and *Actinostachys*) with about 30 species, mostly tropical.

FAMILY LYGODIACEAE

Rhizomes long-creeping, hairy; leaves indeterminate in growth, climbing and often twining, the primary divisions alternate along the elongating stem-like rachis; sporangia often in 2 rows, densely spaced along specialized slender lobes of the ultimate segments, each sporangium covered by an indusium-like flange of tissue; the annulus a subapical ring of thickened cells; 1 genus (*Lygodium*) with about 25 species, mostly distributed in the tropics.

FAMILY ANEMIACEAE

Rhizomes creeping to erect, hairy; leaves with the basal pair (or, rarely, more) of primary divisions modified, mostly lacking laminar tissue, and densely covered with sporangia; sporangia with the annulus, a subapical ring of thickened cells; 1 genus (*Anemia*) with about 100 species, mostly in the Neotropics.

ORDER SALVINIALES

FAMILY SALVINIACEAE (Floating ferns)

Plants heterosporous; stems usually relatively short, mostly appearing dichotomously branched, sometimes lacking roots; leaves either alternate and 2-lobed with 1 lobe green and floating and the other submerged and white or translucent (*Azolla*) or in whorls of 3 with 2 leaves unlobed and floating and the 3rd submerged, modified, and appearing similar to a mass of dichotomously branching roots (*Salvinia*); sori from the submerged lobe or leaf, enclosed in a globose indusium, each containing either 1 megaspore or several microspores; 2 genera, sometimes treated as separate families (Azollaceae and Salviniaceae), *Azolla* (about 6 species) and *Salvinia* (about 10 species), of floating aquatics, distributed nearly worldwide but most diverse in the tropics.

FAMILY MARSILEACEAE (Clover ferns)

Plants heterosporous; rhizomes long-creeping, slender, glabrous or hairy; leaves with 2 or 4 leaflets at the petiole tip or lacking a blade altogether, the venation of leaflets dichotomously branching; sori enclosed in stalked bean-shaped sporocarps (highly modified leaves), very complex internally, each containing both megasporangia and microsporangia; 3 genera of mostly aquatic plants rooted in the substrate—*Marsilea* (waterclover), *Pilularia* (pillwort), and *Regnellidium* with about 75 species found nearly worldwide.

ORDER CYATHEALES (Tree ferns)

FAMILY CYATHEACEAE (Scaly tree ferns)

Stems erect and mostly trunk-like (to 25 metres, [82 feet]) or less commonly creeping or sprawling to short-ascending, scaly near the tip (sometimes also hairy) and usually with a mantle of roots; leaves mostly large (up to 5 metres [about 16 feet]), but in a few species only 10–40 cm (about 4–16 inches), 1 to 4 times pinnately compound (rarely entire), the segments sometimes deeply lobed, scaly, at least on the petioles, sometimes also hairy or spiny; sori in various positions on the leaf undersurface, the sporangia on a usually short receptacle, the indusium absent or varying from deeply 2-lobed to saucer-shaped, cup-shaped, or globose, the annulus slightly to moderately oblique; spores variously ornamented on the surface, often with an equatorial flange or girdle; 3 genera of tree ferns (*Alsophila*, *Cyathea*, and *Sphaeropteris*) with more than 600 modern species, widely distributed in tropical regions.

FAMILY THYRSOPTERIDACEAE

Stems erect and trunk-like or sprawling, hairy and with a mantle of roots, often producing slender runners; leaves large (up to 3.5 metres [11.5 feet]), 3 to 5 times pinnately compound, the axes grooved on the upper surface, the fertile portions with reduced laminar tissue; sori situated in cup-like indusia; sporangia attached to a club-shaped receptacle, the annulus oblique; spores 3-angled, the surface granular; 1 genus (*Thyrsopteris*) with a single species (*T. elegans*), endemic to the Juan *Fernández Islands*.

FAMILY LOXSOMATACEAE

Rhizomes long-creeping, hairy; leaves medium-sized to large (0.5–5 metres [about 1.5–16 feet] long), 2 times pinnately compound, the segments deeply lobed, bristly hairy; sori marginal on the underside of the segments, the sporangia on an elongated receptacle that is often partially exserted from the more or less urn-shaped indusium, the annulus slightly oblique; spores often strongly ornamented on the surface but lacking an equatorial flange or girdle; 2 genera, *Loxsoma*, with 1 species (*L. cunninghamii*) in northern New Zealand, and *Loxsomopsis*, with 1 species (*L. pearcei*) from Costa Rica to Bolivia.

FAMILY CULCITACEAE

Stems variously prostrate and creeping or loosely ascending but usually not trunk-like, hairy, sometimes with a mantle of roots; leaves large (up to 3 metres [almost

10 feet]), 4 to 5 times pinnately compound; sori marginal on the underside of the segments, the indusia unequally 2-lobed or of 2 separate flaps, more or less cup-shaped, the sporangia mixed with paraphyses, the annulus slightly oblique; spores 3-angled, the surface finely pitted or wrinkled to nearly smooth; 1 genus (*Culcita*) with 2 or more species that are widely distributed, mostly in tropical regions.

FAMILY PLAGIOGYRIACEAE

Rhizomes creeping or, more commonly, erect, the tip (and young leaves) covered with mucilage from secretory hairs; leaves 1 time pinnately compound, the petiole bases swollen, dimorphic, the fertile fronds contracted and bearing dense sporangia on the undersurface; the annulus slightly oblique; spores 3-angled, the surface usually with coarse tubercles; 1 genus (*Plagiogyria*) with about 15 species, distributed in tropical regions.

FAMILY CIBOTIACEAE

Rhizomes massive, creeping to erect and often trunk-like (up to 6 metres [almost 20 feet]), with soft yellow hairs toward the tip; leaves large (up to 4 metres [13 feet]), 2 or 3 times pinnately compound, the segments often deeply lobed; sori marginal, the indusium of 2 unequal flaps fused basally, the sporangia sometimes mixed with paraphyses, the annulus slightly oblique; spores 3-angled, the surface slightly to strongly ridged, also with a prominent equatorial flange or girdle; 1 genus (*Cibotium*) with about 11 species, distributed in tropical regions.

FAMILY DICKSONIACEAE (Hairy tree ferns)

Stems mostly erect and trunk-like (up to 10 metres [about 33 feet]) or, less commonly, smaller, hairy near the tip and usually with a mantle of roots; leaves mostly large (up to 3.5 metres [11.5 feet]), 2 to 3 times pinnately compound, the segments sometimes deeply lobed, hairy, at least on the petioles; sori in various positions on the leaf undersurface, the sporangia on an usually short receptacle, often mixed with paraphyses, the indusium absent or unequally 2-valved to cup-shaped, the annulus slightly to moderately oblique; spores variously and usually strongly ornamented on the surface but lacking an equatorial flange or girdle; 3 genera (*Calochlaena, Dicksonia,* and *Lophosoria*) with about 30 modern species, widely distributed in tropical regions but not occurring natively in Africa.

FAMILY METAXYACEAE

Rhizomes not trunk-like, short-creeping or ascending, somewhat flattened, hairy, at least near the tip, the relatively dense roots not forming a mantle; leaves 1–2 metres (3.3–6.6 feet) long, 1 time pinnately compound, the leaflets unlobed, sori scattered on the undersurface of the leaflets, round, lacking an indusium, the sporangia mixed with paraphyses, the annulus slightly oblique; spores globose, finely sculptured, lacking an equatorial flange or girdle; 1 genus and 2 species (*Metaxya rostrata* and *M. lanosa*), of low elevations in the Neotropics, particularly the Amazonian region.

ORDER POLYPODIALES (Known as Filicales in some older literature)

SUBORDER LINDSAEINEAE

FAMILY LINDSAEACEAE (Lace ferns)

Plants mostly in soil or on rocks; rhizomes short- to long-creeping, hairy or scaly; leaves 1 to 3 times pinnately compound, usually glabrous, sori marginal or submarginal, the indusium either lateral and opening toward the margin or the sori protected by a reflexed segment margin; sporangia with the annulus vertical; spores mostly trilete (tetrahedral-globose); about 7 genera and 200 species, distributed in tropical regions.

FAMILY CYSTODINACEAE

Plants similar in appearance to a small tree fern; 1 genus with a single species (*Cystodium sorbifolium*); distributed in lowland rainforests in insular Southeast Asia.

FAMILY LONCHITIDACEAE

Plants have a Neotropical distribution; 1 genus (*Lonchitis*) with about 30 species.

SUBORDER SACCOLOMATINEAE

FAMILY SACCOLOMATACEAE

Plants mostly in soil; rhizomes short-creeping to erect, sometimes appearing as short trunks, scaly; leaves 1 to 5 times pinnately compound, mostly glabrous; sori marginal or submarginal, the indusium pouch-like, opening toward the margin; sporangia with the annulus vertical; spores trilete (tetrahedral-globose); 1 genus with about 12 species, distributed in tropical regions.

SUBORDER DENNSTAEDTIINEAE

FAMILY DENNSTAEDTIACEAE (Cup ferns, bracken)

Plants mostly in soil, occasionally climbing; rhizomes mostly very long-creeping (to more than 100 metres [330 feet], in *Pteridium*), hairy; leaves 2 to 4 times pinnately compound, glabrous or hairy; sori mostly marginal or submarginal, discrete or in a more or less uninterrupted line, the indusium cup-shaped or lateral and elongate, sometimes also with the segment margin reflexed; sporangia with the annulus vertical; spores mostly trilete (tetrahedral-globose); about 11 genera with about 160 species, distributed nearly worldwide but most diverse in tropical regions.

SUBORDER PTERIDINEAE

FAMILY PTERIDACEAE

Plants in soil, on rocks, epiphytic or aquatic; rhizomes short- to long-creeping or erect, usually scaly; leaves entire to highly divided, pinnately or, less commonly, palmately or pedately, occasionally the vegetative and fertile leaves dimorphic (the fertile ones with reduced laminar tissue), glabrous or, more commonly, with hairs, scales, or farina (powdery white or coloured deposit); sori variously positioned, discrete, in lines or bands, or, less commonly, covering the undersurface, the indusium usually absent, but in marginal sori the segment margin often reflexed; sporangia with the annulus vertical; spores trilete (tetrahedral-globose) or, uncommonly, monolete (more or less bean-shaped); about 50 genera with about 950 species, distributed nearly worldwide. The family is quite variable morphologically and has been divided into as many as 5–8 subfamilies.

SUBORDER ASPLENIINEAE

FAMILY ASPLENIACEAE

Plants in soil, on rocks, or epiphytic; rhizomes short- to long-creeping or erect, usually scaly, the scales usually clathrate (the cells with dark adjoining walls and clear lateral walls); leaves entire or lobed to highly pinnately divided, rarely dichotomously divided, glabrous or with inconspicuous hairs, rarely scaly; sori elongate along the veins, oblong to linear, usually discrete, the indusium usually lateral; sporangia with the annulus vertical; spores monolete (more or less bean-shaped); 1–10 genera (depending on how much *Asplenium* is divided) with about 800 species, distributed nearly worldwide.

FAMILY ATHYRIACEAE (Lady ferns)

Plants in soil or on rocks; rhizomes short- to long-creeping or erect, scaly; leaves 1 to 4 times pinnately divided, glabrous, hairy, glandular, or occasionally scaly; sori round or elongate along the veins, sometimes back to back along a vein or curved around a vein ending (J-shaped), the indusia absent or variously linear to oblong, kidney-shaped, cup-shaped, or dissected into slender filaments; sporangia with the annulus vertical; spores monolete (more or less bean-shaped); 2 genera (about 180 in the genus *Athyrium*).

FAMILY BLECHNACEAE (Chain ferns)

Plants in soil or on rocks, less commonly epiphytic, rarely climbing; rhizomes short- to long-creeping or erect (occasionally trunk-like), scaly; leaves 1 time pinnately compound or lobed, less commonly 2 times pinnately compound, rarely entire, sometimes the vegetative and fertile leaves strongly dimorphic (the fertile ones with reduced laminar tissue), glabrous or, less commonly, hairy; sori oblong to elongate, often attached along one side of the areole formed by a network of veins, often appearing as a band or chain on each side of the segment midrib, the indusia lateral and linear to oblong, opening toward the midrib; sporangia with the annulus vertical; spores monolete (more or less bean-shaped); about 24 genera with about 265 species (about 150 in the largest genus, *Blechnum*), distributed nearly worldwide but most diverse in tropical regions.

FAMILY CYSTOPTERIDACEAE

Plants are small or medium-sized; leaves are thin; sori are small, round, and naked; 3 genera (*Acystopteris*, *Cystopteris*, and *Gymnocarpium*).

FAMILY DESMOPHLEBIACEAE

Plants have a thickened vein inside the edge of each compound leaflet; 1 genus (*Desmophlebium*) with 2 species.

FAMILY DIPLAZIOPSIDACEAE

Plants medium to large in soil or on rocks; rhizomes short- to long-creeping or erect, scaly; leaves pinnately divided, glabrous, hairy, glandular, or occasionally scaly; sori round or elongate along the veins; sporangia with the annulus vertical; spores monolete (more or less bean-shaped); 2 genera (*Diplaziopsis* and the monotypic *Homalosorus*).

FAMILY HEMIDICTYACEAE

Plants with a Neotropical distribution; veins are netted only halfway each leaflet across the pinnules, family consists of a single species (*Hemidictyum marginatum*).

FAMILY ONOCLEACEAE

Plants in soil; rhizomes short- to long-creeping or erect (occasionally trunk-like in *Onocleopsis*), scaly, sometimes with runners; leaves strongly dimorphic, the vegetative leaves one time pinnately compound, the leaflets sometimes deeply lobed, the fertile leaves one to three times pinnately compound, with contracted laminar tissue, the margins of the segments rolling inward at maturity to enclose the sporangia, the whole leaf often becoming a brown and hardened overwintering resting structure at maturity (except in *Onocleopsis*), glabrous or hairy; sori round, the indusia asymmetrically cup-shaped to more or less globose, rarely absent; sporangia with the annulus vertical; spores monolete (more or less bean-shaped to globose), green; 4 genera with 5 species, distributed mostly in temperate regions (Neotropical in *Onocleopsis*).

FAMILY RHACHIDOSORACEAE

1 genus (*Rhachidosorus*) with about 7 species.

FAMILY THELYPTERIDACEAE

Plants in soil or, less commonly, on rocks; rhizomes short- to long-creeping or erect, scaly; leaves mostly 1 or 2 times pinnately divided, rarely highly divided, most commonly with slender needle-like hairs (these sometimes also on rhizomes, indusia, or sporangia), occasionally also with scattered scales or glabrous; sori round or elongate along the veins, the indusia absent or kidney-shaped; sporangia with the annulus vertical; spores monolete (more or less bean-shaped); 30 genera (depending on how much *Thelypteris* is divided) with 900 species, found nearly worldwide.

FAMILY WOODSIACEAE

Plants in soil or on rocks; rhizomes short- to long-creeping or erect, scaly; leaves 1 to 4 times pinnately divided, glabrous, hairy, glandular, or occasionally scaly; sori round or elongate along the veins, sometimes back to back along a vein or curved around a vein ending (J-shaped), the indusia absent or variously linear to oblong, kidney-shaped, cup-shaped, or dissected into slender filaments; sporangia with the annulus vertical; spores monolete (more or less bean-shaped); about 1 genus (Woodsia) with about 40 species.

SUBORDER POLYPODIINEAE

FAMILY DAVALLIACEAE (Rabbit's-foot fern)

Plants epiphytic or sometimes on rocks; rhizomes long-creeping, often somewhat flattened, scaly; leaves 1 to 4 times pinnately divided, rarely undivided, the petioles jointed at their bases, the blades glabrous or, less commonly, hairy; sori round, rarely elongate along the segment margin, the indusia cup-shaped or kidney-shaped, rarely elongate; sporangia mixed with hairlike paraphyses, the annulus vertical; spores monolete (more or less bean-shaped); 1 genus (*Davallia*) with about 40 species, distributed in tropical and warm temperate regions.

FAMILY DIDYMOCHLAENACEAE

1 genus with a single species (*Didymochlaena truncatula*).

FAMILY DRYOPTERIDACEAE

Plants in soil, on rocks, or epiphytic; rhizomes short- to long-creeping or ascending to erect, scaly; leaves entire or 1 to 4 times pinnately divided, the vegetative and fertile leaves or leaflets occasionally dimorphic, glabrous, hairy, glandular, or scaly; sori round or densely covering the undersurface, the indusia absent or umbrella-shaped to kidney-shaped; sporangia with the annulus vertical; spores monolete (more or less bean-shaped); 45 genera with about 1,700 species, the largest genera, *Dryopteris* (log fern, about 250 species), *Polystichum* (shield fern, about 250 species), and *Elaphoglossum* (tongue fern, 600–700 species), distributed nearly worldwide.

FAMILY HYPODEMATIACEAE

Small family of 2 genera, *Hypodematium* (about 20 species) and *Leucostegia* (2 species).

FAMILY LOMARIOPSIDACEAE

Plants in soil, on rocks, or climbing (hemiepiphytic); rhizomes short- to long-creeping, sometimes with runners, scaly; leaves mostly 1 time pinnately compound, the leaflets often jointed to the axis, glabrous, hairy, or scaly; sori round, the indusia absent or kidney-shaped; sporangia with the annulus vertical; spores monolete (more or less bean-shaped); 4 genera (*Cyclopeltis, Dracoglossum, Dryopolystichum,* and *Lomariopsis*) with about 70 species, widely distributed in tropical regions.

FAMILY NEPHROLEPIDACEAE

Plants widely distributed in tropical regions and sometimes cultivated as ornamental sword ferns; 1 genus (*Nephrolepis*) with about 30 species; formerly placed in Lomariopsidaceae.

FAMILY OLEANDRACEAE

Plants in soil, on rock, or climbing (hemiepiphytic); rhizomes long-creeping, sometimes with elongate, ascending, or climbing branches, scaly; leaves undivided, the petioles jointed to short, persistent stubs (phyllopodia) along the rhizomes, the blades scaly; sori round, the indusia kidney-shaped to nearly U-shaped; sporangia mixed with hairlike paraphyses, the annulus vertical; spores monolete (more or less bean-shaped); 1 genus (*Oleandra*) with about 40 species, widely distributed, mostly in tropical regions.

FAMILY POLYPODIACEAE (Polypodies)

Plants epiphytic, on rock, or occasionally in soil; rhizomes mostly long-creeping, sometimes somewhat flattened, scaly, the scales often clathrate (the cells with dark adjoining walls and clear lateral walls); leaves undivided or 1 time pinnately lobed or compound, rarely more divided, sometimes the vegetative and fertile leaves dimorphic, the petioles often jointed to short, persistent stubs (phyllopodia) along the rhizomes, the blades glabrous, hairy, or scaly; sori round or oblong to elliptic, rarely elongate or covering the undersurface, lacking indusia, but sometimes covered with overlapping scales when young; sporangia sometimes mixed with paraphyses, the annulus vertical; spores monolete (more or less bean-shaped) and often golden yellow, or trilete (more or less globose) and green; about 65 genera, including *Polypodium* (polypody), *Lepisorus* and *Pleopeltis* (scaly polypodies), *Phlebodium* (hare's foot fern), *Campyloneurum* (strap fern), *Microgramma* (vine fern), *Pyrrosia* (felt fern), *Drynaria* (oak-leaf fern), and *Platycerium* (staghorn fern), with about 1,650 total species. *Grammitis* and some 3–15 segregate genera (about 700 species) with green trilete spores and often characteristic dark hairs formerly classified in Grammitidaceae are now considered a specialized subgroup of Polypodiaceae.

FAMILY TECTARIACEAE

Plants in soil or on rocks; rhizomes mostly short-creeping or ascending, scaly; leaves undivided or 1 or 2 times pinnately compound, rarely highly divided, glabrous or hairy, the hairs often appearing minute and jointed; sori round, the indusia absent, umbrella-shaped to kidney-shaped, or dissected into slender filaments; sporangia with the annulus vertical; spores monolete (more or less bean-shaped); 7genera with about 230 species widely distributed in tropical regions.

6

CHAPTER

PSILOTUM

Polypodiopsida (after Schuettpelz et al., 2016)

[MONILOPHYTES] (Chase & Reveal, 2009)

SUBCLASS PSILOTIDAE (Chase & Reveal, 2009)

[FERNS] (Christenhusz & Chase, 2014)

CLASS POLYPODIOPSIDA (**PPG 1, 2016** = Psilotopsida after Smith et al., 2006)

SUBCLASS OPHIOGLOSSIDAE (Christenhusz & Chase, 2014; **PPG 1, 2016**)

PSILOTALES (Smith et al., 2006, Christenhusz & Chase, 2014; **PPG 1, 2016**)

PSILOTACEAE (Smith et al., 2006, Christenhusz & Chase, 2014; **PPG 1, 2016**)

The whisk ferns are morphologically and anatomically simple in construction and therefore evolutionary biologists hypothesize *Psilotum* to be an evolutionary throwback to evolution of vascular plants. For several years when physical attributes were considered, the genus was believed to bear a lot in common with extinct Rhyniophytes which arose during early Devonian period. However, with genomics being studied to solve evolutionary and phylogenetic puzzles, molecular analysis has brought *Psilotum* close to Ophioglossales and in recent classifications, is kept within fern lineage. It is also debated that perhaps whisk ferns represent a reduction from a traditional fern.

PSILOTALES – The living fossil

PSILOTACEAE — Whisk ferns; incl. Tmesipteridaceae. Two genera (*Psilotum, Tmesipteris*), ca. 12 total spp. (2 in *Psilotum*); monophyletic (Hasebe et al., 1995; Pryer et al., 2001a, 2004).

Roots are absent with stems bearing reduced, unveined or single veined euphylls; sporangia are large, with walls two cells thick, lacking an annulus; two or three sporangia fused to form a synangium, seemingly borne on the adaxial side of a forked leaf; spores reniform, monolete, many (> 1000) per sporangium; gametophytes subterranean (*Psilotum*), non-photosynthetic, mycorrhizal; $x = 52$.

---

*PPG1 Schuettpelz et al., (2016)

PSILOTUM

Psilotum has a wide distribution in both tropical and subtropical land masses. It may abound certain areas but in certain habitats it is considered rare or even endangered. It consists of two common species, *P. nudum* and *P. flaccidum,* however, some workers have also reported *P. triquetrum* and *P. complanatum*. The genus is pantropical and subtropical in distribution, while *Psilotum complanatum* occurs as epiphyte on tree ferns, on coconut trunks or at the base of trees. Occasionally members may be terrestrial, growing in soil or among exposed rocks. *Psilotum nudum* grows remarkably well under greenhouse conditions and is cultivated in most botanical gardens, in the temperate regions.

SPOROPHYTE

Morphology

Depending upon its habitat, *P. nudum* may be erect or pendent as in *P. flaccidum*, short or reach a height of 75-100 cms. The plant body consists of a basal branched rhizome system which generally remains hidden beneath the soil and a slender, upright, green aerial portion which is freely and dichotomously branched and bears small appendages and synangia (fused sporangia) at the time of reproduction **(PLATE 3A)**.

- The much-branched rhizome bears no leaves or roots and is much softer than the aerial stem. The surface is generally covered with fine, brown hairs, which may be better developed in some regions than in others. Prantl states that the branching of *Psilotum* cannot be regarded as dichotomous, as one of the branches arises in the axil of a leaf while the other continues the growth and checks the phyllotaxis of the main shoot. Roots are absent. The branched rhizome system, which bears numerous scales grows by means of apical meristems located at tips of ultimate branches and anchors the plant in the substratum. It also serves as the absorptive surface. A mycorrhizal intracellular fungus related intimately to the physiology of the plant, infecting through rhizoids is present in cells of the outer cortex.
- Any one of the rhizome tips may turn upwards and by apical growth produce an aerial branch system (Fig. 6.1A). The basal part of the shoot may be cylindrical, with longitudinal ribs whereas the more distal aerial stems are provided with three longitudinal ridges.
- The aerial shoots bear minute scale-like lateral appendages without constant and definite arrangement in the lower stouter region. Leaves better called appendages are bi-lobed, a feature which makes it easy to locate even very young synangia which are reproductive structures borne in these axils of the fertile plant (Fig. 6.1B).
- The synangia themselves are for the most part tri- locular, but bilocular ones are occasionally found (**PLATE 3B**).

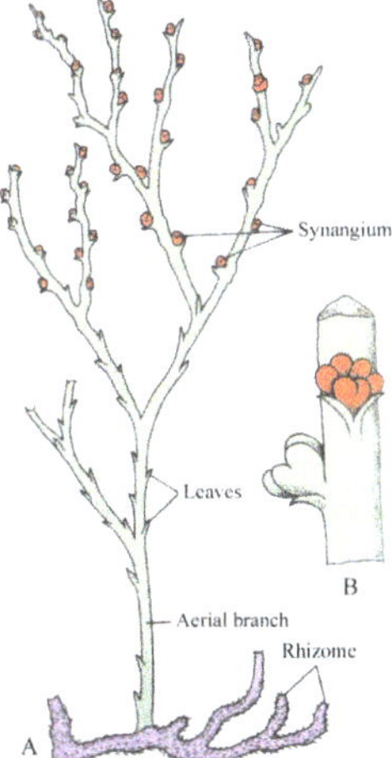

Figure 6.1A, B: *Psilotum* sp. (Morphology)-Diagrammatic Sketch

A - Is a mature sporophyte with subterranean rhizome, aerial stem with dichotomous branching. B - Is a closer view of stem portion showing appendages and synangia in the axils.

Anatomy

Stem

The aerial stem forms externally the most noticeable part of a *Psilotum* plant. At its base, near the soil, stem is smooth, brown, and circular in outline, but higher up it gradually becomes conspicuously ribbed and green in colour. The apical meristem of rhizomes and aerial branches has a single, large, wedge-shaped apical cell which divides repeatedly giving rise to additional meristematic cells differentiating eventually into tissues comprising the three primary tissue systems.

- The stem is ribbed in the stouter parts of the plant while towards the apex it is triangular in outline.
- The outermost layer, epidermis is cuticularized, and fairly large number of stomata are present in the grooves of the stem between the ridges. The guard-cells which are slightly sunk below the level of the other epidermal cells, have cuticularized outer walls.
- The cortex is composed of three zones, an outer, parenchymatous and assimilating, a middle sclerenchymatous, and an inner parenchymatous surrounding the stele.
- The outer parenchymatous zone is composed of peculiar cells, whose lateral walls bulge out at regular intervals, and each cell has 2-4 such swellings. A conspicuous row of intercellular spaces is formed when the swellings of adjacent cells touch each other. The cells themselves are nucleated and contain chlorophyll.

- The sclerenchymatous zone consists of thick-walled fibrous elements with pointed ends, and the walls are perforated by small slit-like pits.
- Towards the centre, this sclerenchymatous zone passes gradually into parenchymatous region immediately surrounding the stele (**PLATE 3C**). The walls of the cells are perforated by numerous oval or circular pits, which vary in size. Starch is generally found in the cells, besides conspicuous silica-rich nodules of varying shape and size in the lower and older parts of the plant.
- The endodermis, is a badly defined layer, and the thickenings on the radial walls stand out very distinctly.
- The major part of the aerial branches except the ultimate branches and above the basal region has siphonostele. The pith is sclerotic. The centre of the stele is occupied by a group of sclerenchymatous fibres with numerous, small, simple pits. In the stouter parts of the plant the fibres form a conspicuous group; higher up the number gradually decreases until only two or three, and finally none at all, are present. The star-like xylem surrounds the central group of fibres. The projecting points of the star-like xylem mark the position of the protoxylem-groups, the number of which varies with size and region of the stem. In stouter branches nine or ten such masses may be found, while in the small apical branches two or three only may be present. In the latter case the xylem may be present as a small central strand, or it may form two small masses lying close together, each containing one group of protoxylem. The metaxylem is made up of ordinary scalariform and pitted tracheids, the protoxylem of spiral elements (Fig. 6.2A).
- Between the endodermis and the xylem, a mass of tissue, very difficult to differentiate is found.
- There is no clearly defined pericyclé, the tissue consisting of a mass of parenchymatous elements of varying size is seen. Treatment with aniline chloride indicates lignification in some of the cells, especially those at the corners; and this is more prominent in the tissue lying outside the projecting tips of protoxylem. In longitudinal sections the mass of the tissue is seen to be composed of elongated parenchymatous cells with conspicuous nuclei and numerous small oval or circular pits or pores.
- Besides the nucleated parenchyma, long tube-like elements occur singly, or occasionally one or two together, which may possibly be regarded as sieve- tubes.

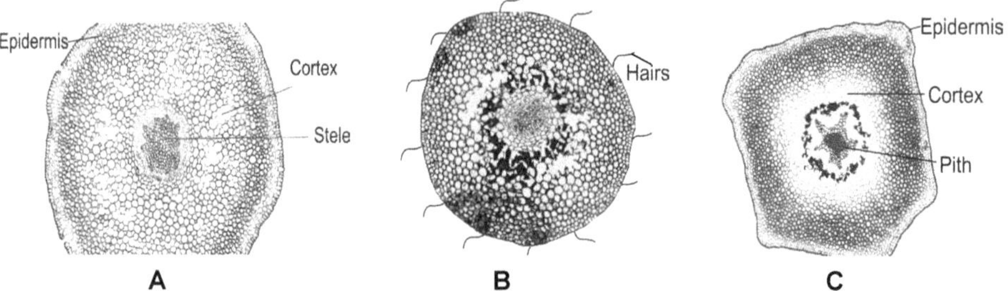

Figure 6.2 A-C: *Psilotum* (Anatomy)

A -T.S. Aerial stem to show epidermis followed by cortex which is differentiated into three zones, a chlorenchymatous zone that acts as a photosynthetic tissue in absence of leaves, a sclerenchymatous and a broad parenchymatous zone. Endodermis and pericycle are not well-defined while major part of the stem has siphonostele with sclerotic pith and stellate xylem. B - T.S. Rhizome shows a less differentiated organization, the epidermis from which arise rhizoids, is followed by parenchymatous ground tissue or cortex, the outermost few layers of which are dark due to phlobaphene. There is also presence of fungal mycelia in a few cells below the epidermis. The vascular system is represented by protostele and xylem occasionally appears diarch. C- T.S. Stem passing through the upper region, the outline is angular and the stele is siphonostele.

Rhizome

- The outermost tissue epidermis, remains ill-defined and is followed by a massive cortex (**PLATE 3D**).
- The main mass of the cortex comprises large thin-walled cells containing starch. The two or three layers, however, next to the endodermis, possess walls coloured a dark brown. This coloration is due to * gélification ' and ' humifaction ' phlobaphene in the cell wall (Fig. 6.2B). Many of the cells of outer thin-walled cortex are crowded with filaments of the fungus which shows globular or pear-shaped swellings on the mycelial threads which presumably are the aborted sporangia or 'sporangoïds'. The cells of the outermost cortical layer grow out into absorptive hairs where each hair is composed of a basal cell and the main mass of the hair itself.
- The typical structure of stele in a simple branch consists of a diarch band of xylem. At two extremities of this band are the protoxylem elements with spiral thickenings. However, the xylem-strand exhibits considerable variation in size and shape. While in smaller branches, only two or three tracheids may be found, in stouter and more mature regions of the rhizome, an irregular mass or a band-shaped strand, or again a central group of xylem which is more or less circular in outline may be found. The tracheids are scalariform, with no parenchyma interposed, and the spiral protoxylem elements are reported to be generally absent.
- Surrounding the xylem is the phloem, similar in general respects to that found in the aerial stem, but is less well developed. The 'sieve-tubes' and nucleated parenchymatous cells are present, but the pitting on the walls is more or less obscure. The lignification of the sieve tubes is also not definite. The endodermis although not so easily distinguishable as in the aerial stem, is also present.

*The rhizome is protostelic, becoming an exarch siphonostele throughout a considerable portion of the aerial branch system, while the upper branches may be strictly actinostelic.

The Intermediate Region

- This region includes the base of the aerial stem, from the level at which the green ribbed surface of the stem changes to the smooth brown cylindrical structure, down to the point of meeting the rhizome. At the extreme base, the stele resembles ordinary rhizome branch, viz. a central strand of xylem of varying size and shape, with no clearly defined protoxylem, and surrounded by phloem tissue, making it a protostele. The walls of the two or three cortical layers next to the endodermis are deeply coloured.
- A little above there are no superficial hairs, and the thin-walled cortex is gradually replaced by the brown- walled tissue. The xylem increases in amount, and interspersed in xylem tracheids lie parenchymatous cells, but the protoxylem is not distinguishable, although further up the stem two and then three or more groups of protoxylem tracheids may be seen. The secondary tracheids, appear in this region. Higher up, the stele passes through an interesting stage leading to siphonostele (Fig. 6.2C). The amount of xylem not only increases in extent, but the three or four protoxylem groups become readily distinguished, and the parenchyma, instead of being scattered amongst the tracheids, lies in the centre as a more or less distinct pith.
- Further up again, the xylem forms a complete, unbroken ring around the central pith. This tissue is composed of nucleated elements, with thin areas on their walls. No sieve-tubes or phloem has been observed, as also the presence of endodermis is missing. A few sections higher up, the organization of aerial stem becomes visible - the first of the central fibres makes its appearance in the pith, the number then gradually increases, until the typical central mass is formed - siphonostele, surrounded by the xylem with radiating protoxylem. The cortex gradually changes from the uniform mass of brown-walled cells to the usual arrangement of parenchymatous and sclerenchymatous zones found in the aerial stem.

Leaf

- The foliar appendages are small scale-like structures which are irregularly distributed on the aerial stem.
- Internally the appendage has outermost epidermis which lacks stomata but is cuticularized.
- The epidermis is followed by a region that consists of photosynthetic parenchyma cells which are continuous, lower down, with similar tissue of the stem.
- There is no vascular bundle in the appendage of *P. nudum* although in *P. flaccidum* a leaf trace ends at the base of foliar structure.
- Grouped generally on the upper part of the stems, are bilobed appendages, each of which is associated with a three lobed sporangium or synangium.

Reproduction

Sporangium/synangium

- The sporangia better known as synangia occur, as a rule, on the later-formed branches, but they may also be found on older branches near the base of the stem.
- The synangium in the mature condition is generally a three lobed structure, 2-3 mm wide and is associated with a forked foliar appendage (Fig. 6.3A, **PLATE 3E**). Each lobe of the synangium corresponds to an internal spore chamber and exhibits loculicidal dehiscence at maturity. The sporangial sac which shows homosporous condition has a well-defined tapetum around the spore sac (Fig. 6.3B, **PLATE 3F,G**).

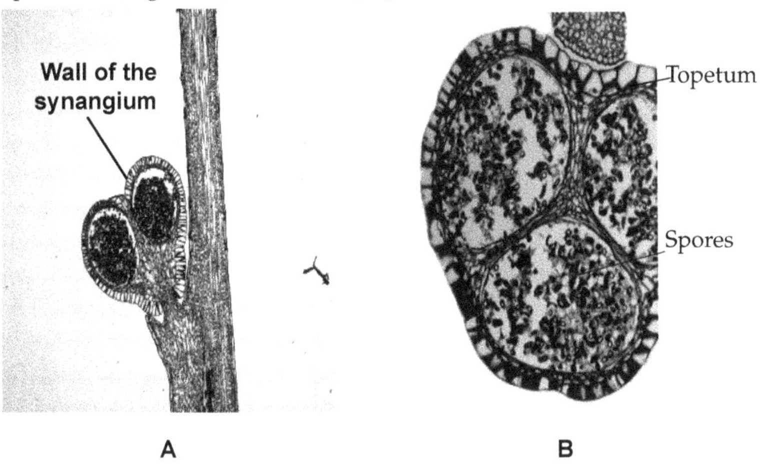

Figure. 6.3 A,B: *Psilotum* sp. (Synangium)

A - L.S. of fertile axis bearing synangia. B - T.S. Synangium showing sporangial wall, distinct tapetum and locule with spores.

- In early stages of development, it is difficult to determine whether a lateral primordium will become a single vegetative appendage or a fertile branch. Bierhorst concluded that original appendage is in fact the sporangial or spore producing apparatus and that the subjascent leaf-like lobe is a lateral outgrowth on the fertile axis. The researches and results of Bower seem to be conclusive indicating that the sporangium is not a terminal outgrowth of a reduced axis which bears two leaves, but it is an outgrowth from the sporangial leaf, which is bi-lobed possibly for a protective purpose. The whole sporangiophore is then a single foliar member.
- The primordium of the fertile axis exhibits apical growth and the sites of future sporangial locules become apparent early in development. In each case, these divisions result in the setting apart of primary wall initials and

primary sporogenous cells (**PLATE 3E-G**). By repeated divisions of primary wall initials, a sporangial wall of four to five layers is produced. Not all derivatives of primary sporogenous cells form sporogenous tissue which give rise to sporocytes as some disintegrate adding to the nourishing fluid.

- Eusporangiate development leads to formation of a large number of spores. Individual spores are bilateraly symmetrical while the sporangial axis is vasculated.

Interpretations

The morphological interpretations of the spore producing structure in *Psilotum* are varied and controversial.

i) It is a foliar structure (Bower's interpretation) and forked leaves are bifid sporophylls each bearing a triclocular fructification (a single partitioned sporangium).

ii) It is a short lateral shoot, the short fertile branch bears two sterile appendages and terminates in a trilocular fructification.

iii) It is an organ or a group of organs *sui generis*.

**The spore producing structure in the Psilotales has been interpreted by some investigators as being a synangium – that is the fusion product of two or more sporangia. Others consider the foliar appendage to a sporophyll bearing a trilocular sporangium. A third school of thought interprets the sporangium as occupying the terminus of a short or lateral branch. According to the last interpretation the sporangium is fundamentally a trilocular structure, cauline in nature and terminal in position. The presence of vascular tissues in the sporangial axis of *Psilotum* tends to support the axial concept of the sporangium. The sporangial axis has been compared with the sporangiophore in Equisetales. The concept that each unit of the fertile axis represents phylogenetically a condensation of a more elaborate branch system may also be true.

Vegetative Reproduction

Gemmae: The brood bodies develop in large numbers on the surface of rhizome between axils of the rhizome branches. The gemmae have an apical cell each, and cells are filled with starch grains. These bodies germinate while still attached to the parent rhizome. Gemmae are also borne on the surface of the prothallus.

GAMETOPHYTE

The spores are slow to germinate and the contents project as a small lobule which subsequently increases in size. An apical cell is organized during development and the derivative mass is penetrated by an endophytic fungus. After some more growth, a gametophyte more commonly called a prothallus is formed.

- The gametophyte ranges from 0.5 mm to 2.0 mm in diameter and is several millimetres in length, grows on trunks of tree ferns or in crevices of rocks, is occasionally subterranean in habitat.
- The color of the mature gametophyte varies from pale yellow to dark brown, or occasionally to nearly black with meristematic margins more conspicuous due to their nearly white color; apices which resemble lateral buds are also common (**PLATE 3H,I**). As the gametophyte is devoid of chlorophyll, it leads a saprophytic mode aided by an endophytic fungus which gains entry through rhizoids.
- The branching pattern in general is quite inconsistent with equal and unequal dichotomies, tricho- tomies and tetrachotomies seen in the prothallus. The angles between the branches vary greatly. This branching is correlated to a considerable degree with apical injury which results in the establishment of lateral axes. The mature gametophytes of *P. nudum* resemble sporophytic rhizome, are more or less cylindrical, radially symmetrical, and invested with rhizoids.
- The gametangia are usually abundant and are distributed uniformly over the surface and length of the mature gametophytes. In some cases where the gametophyte is usually quite large and possesses few rhizoids, the antheridia almost completely cover the surface. At the other extreme, there is an occasional gametophyte on which the antheridia are as much as 4 mm apart. The archegonia are generally not as abundant as the antheridia but are conspicuous dark coloured near the meristematic apices.
- The surface cells of the gametophytes are living at maturity, contain fungal hyphae, and have dark colored, decay resistant, strongly cutinized outer and radial walls, and even the inner corners of the walls are cutinized. The cutinization of the archegonial neck cells, the wall cells of the antheridia, and the cells of the rhizoids is also common. The resistance to decay is borne out by the discovery within the soil of a number of gametophyte "shells" formed by disappearance of internal tissues, leaving only the outer layer, including all of the sex organs and rhizoids well defined. Such phenomenon may assist in fossilization.

Anatomy

- Transverse section is circular in outline, and the surface cells are living at maturity whose outer walls are strongly cutinized (Fig. 6.4A). The ground parenchyma is continuous across the axis of the smaller gametophytes which lack a vascular strand. However, in larger gametophytes which are vascularized, a cortex, composed of thin-walled, living parenchymatous cells, the walls of which are often cutinized and slightly thickened, is formed. These cells also have mycorrhiza and frequently distributed starch grains, with a greater number in the cells near the surface. In the younger portions of the gametophyte of *P. nudum*, the ground parenchyma and surface cells possess scanty cytoplasm, large vacuoles, and a single nucleus.

- The older gametophytes, show multinucleate ground parenchyma and surface cells with somewhat decomposed contents. These cells possess generally one relatively large nucleus and several smaller nuclei. The decomposition of the cell contents has been interpreted by Lawson (1918) to be "the eventual effect of the parasitic nature of the fungus."
- The transition region between a vascularized and a non-vascularized portion of a gametophyte is as follows: First the centrally located tracheids become fewer and as one moves along the gametophyte they become absent. Next, thin-walled elements with starch grains also become fewer in number, followed by the disappearance of endodermis. The vascular strand at this stage is represented by two or three centrally located, thin-walled elements in which fungal hyphae may be present and finally, the distinction between the centrally located cells and the sub-isodiametric cells of the ground parenchyma becomes less significant.
- It is noteworthy that the fungal hyphae, are present besides parenchymatous cells, in the antheridial wall cells, the spermatogenous tissue, the archegonial neck cells, the neck of the archegonium, and in the egg chamber where they often form dense, coiled masses almost completely filling the cell lumen. The rhizoids arise from certain of the superficial cells in the region immediately behind the apex.
- In some of the larger gametophytes, as described by Holloway (1939), a distinct vascular strand is present.
- All cells of prothallus are parenchymatous however, annular and scalariform-reticulate tracheids, surrounded by phloem and endodermis have been shown to occupy the centre of the prothallus. The endodermal cells are somewhat elongate, from 7-12 times as long as wide and possess an essentially unthickened Casparian strip which is finely pitted. All cells of the vascular strand are devoid of both starch and fungi. In such vascularized prothallus, cortex with endophytic mycorrhiza is organized between superficial and vascular tissues (Fig. 6.4B).
- The prothallus which is generally diploid (as most of the sporophytes are tetraploid) possesses antheridia and archegonia on the surface (Fig. 6.4C).
- The striking similarity in general appearance between gametophyte and sporophyte rhizome, coupled with presence of vascular tissue in the abnormal vascularized gametophyte provides additional evidence to the origin of 'Homologous Theory of Alternation of Generations.'

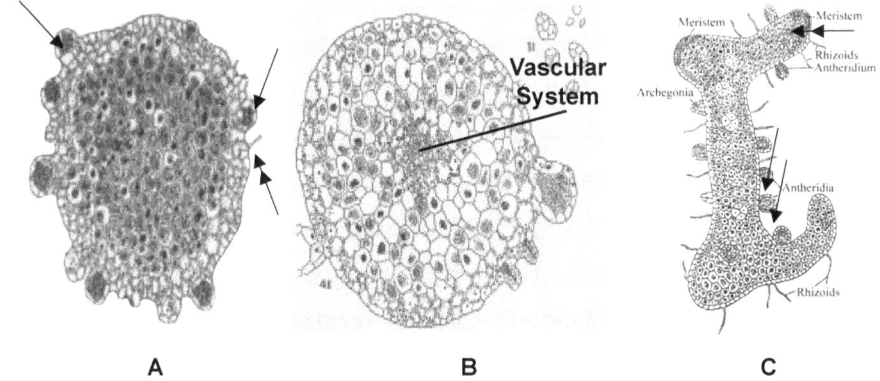

Figure 6.4A-C: *Psilotum* sp. (Prothallus)

A – T.S. passing through a smaller gametophyte which shows antheridia (arrows) that appear to protrude and the ground tissue is parenchymatous. The cortical cells show fungal hyphae and rhizoids (double arrows) are seen to arise from the epidermis. In B – Central cylinder of simple vascular tissue is seen and the outer walls are cutinized. In C – Same in longisection; the distribution of sex organs is quite distinct. Also seen are antheridia (arrow) and archegonia (double arrow).

Sexual Reproduction

- Sex organs are scattered over the surface of the gametophyte and male and female occur with no particular order.
- The sex organs begin development very close to the apex of the gametophyte.
- Antheridial ontogeny is similar to that in eusporangiate ferns, *Lycopodium* and *Equisetum* and the mature antheridium projects above the surface.
- The archegonia also differentiate with a neck and a venter. The archegonial necks are straight and venter is embedded in the gametophyte. With the sloughing off of the upper portion of the neck and disintegration of the axial row, a passageway is created for the entrance of the motile antherozoids.

Young Sporophyte

- The first division in the zygote is by a wall at right angles to the long axis of the archegonium.
- The lower cell is hypobasal while the one towards the neck is epibasal.
- The embryo shows **exoscopic** polarity. The epibasal cell organizes shoot system while the hypobasal cell produces a foot.
- The shoot organizes an apical cell and the foot enlarges sending haustorial outgrowths into the gametophytic tissue and helps in anchorage and absorption.
- Ultimately the actively growing shoot end along with rhizome becomes detached from the foot and the gametophyte through a separation layer and

the apical cell begins to function forming a branch which becomes infected with fungus.

Theoretical Conclusions

- *Psilotum* is generally regarded as representing an ancient type, which has retained to a certain extent some of its primitive features, whilst showing at the same time modifications due to its existence on land.
- As each aerial branch arises from the rhizome and turns to grow vertically upwards above the soil, cortical as well as stelar tissues undergo considerable change. As more mechanical support is needed by the sporophyte, whole brown-walled cortex provides this strength.
- Anatomically, however, the Psilotaceae show certain Lycopodinean features. There is resemblance between the central stele of the Psilotaceae and that found in the axis of *Lepidostrobus brownii*, a lycopsid from Devonian period.
- Four chief points of resemblance, have been cited as the crenulated margin of xylem, the definite layer of endodermis as in *Psilotum*, the slight bulk of phloem tissue and absence of distinctive characters – sieve tubes in it, and lastly, the presence of a parenchymatous pith in *Lepidostrobus* which resembles structures in *Tmesipteris*, though its place is taken by thick- walled sclerenchyma in *Psilotum*.
- There is again a strong resemblance between the aerial and intermediate regions of the stem of *Psilotum* and the fossil stems of *Lepidodendron mundum*. Some preserved sections of the latter show a distinct medulla of varying size, the xylem that forms a definite ring which may be broken at some levels. In a few sections this parenchymatous pith seems to be replaced by thick-walled tissue as in *Psilotum*.
- There also exists affinity of Psilotaceae with Sphenophyllales based exarch condition of xylem. From the anatomical point of view, the characteristic triangular mass of primary wood, which is noticeable in the stem of *Sphenophyllum*, recalls the structure of the smaller branches of an aerial stem of *Psilotum*.
- A strong resemblance exists between the Psilotaceae and the Sphenophyllales in regard to the sporangial structures. Though the two groups are regarded as being related, but probably they represent two distinct lines of evolution. Scott has summarized the situation by suggesting that the Psilotaceae may be allied, though very remotely, to the Lycopodieae, and has branched off from the main line of Lycopod descent very far back. At this point of branching, some characters common to the Sphenophyllales were still retained; and therefore observations of the gametophyte and of the young sporophyte are required that might throw fresh light on this possible relationship. As matters stand, the affinity of the Psilotaceae with the Sphenophyllales is clearly the most marked, but until a further knowledge of the development of the living genera of *Psilotum* and *Tmesipteris* is known, it would seem a little premature to assign to them a definite position in a group which

hitherto has only included fossil forms. In regard to the rhizome of *Psilotum*, a considerable similarity in structure has been found to exist with the fossil root or rhizome.

DO YOU KNOW?

Paleobotanical studies have long demonstrated that the genus *Equisetum* is the sole surviving representative of an extremely ancient and diverse group, the Sphenopsida with at least 60 genera and a number of species. It has a long evolutionary history that can be traced in time as far back as the Late Devonian. Sphenopsids comprising two orders, the Sphenophyllales and the Equisetales, reached a peak in species richness during the Carboniferous period, which occupied paleotropical swamp forest ecosystems. The molecular data indicates that among the extant vascular plants, *Equisetum* (Sphenopsids), *Psilotum* and ferns, and seed plants are more closely related to each other than the relatedness with lycopsids which are monophyletic. This group of tracheophytes has been termed euphyllophytes (Kenrick & Crane, 1997).

Class	POLYPODIOPSIDA
Subclass	OPHIOGLOSSIDAE
Order	PSILOTALES
Family	PSILOTACEAE
	PSILOTUM

QUESTIONS

Q1. *Psilotum* is an evolutionary throwback to evolution. Comment.

Q2. Anatomy of *Psilotum* stem shows interesting transitions in stele of aerial shoot. Discuss.

Q3. Compare the sporangia of *Equisetum* and *Psilotum*.

EQUISETUM

Polypodiopsida (after Schuettpelz et al., 2016)

[MONILOPHYTES] (Chase & Reveal, 2009)

[FERNS] (Christenhusz & Chase, 2014)

CLASS EQUISETOPSIDA = SPHENOPSIDA (Smith et al., 2006 = Polypodiopsida, **PPG 1, 2016**)

SUBCLASS EQUISETIDAE (Chase & Reveal, 2009; Christenhusz & Chase, 2014; **PPG 1, 2016**)

ORDER EQUISETALES (Smith et al., 2006., Christenhusz & Chase, 2014; **PPG 1, 2016**)

FAMILY EQUISETACEAE (Smith et al., 2006; Christenhusz & Chase, 2014; **PPG 1, 2016**)

EQUISETOPSIDA = SPHENOPSIDA (HORSETAILS AND SCOURING RUSHES)

Horsetails are unique survivors of a very ancient group of vascular plants, frequently referred to as the Sphenophyta (Arthrophyta), with a history dating back to the Upper Devonian. *Equisetum*, the extant horsetail has a link in the Cretaceous and possibly as far back as the Triassic (Hauke, 1978) period. *Equisetum* may said to be the oldest living genus of vascular plants (Hauke, 1963) with a great diversity expressed in the early Carboniferous (Fig. 7.1A,B). Despite the striking conservatism of *Equisetum*, the only extant member, architecture, anatomy, and the small number of species (15 and many inter-specific hybrids) in the modern flora, the ability of the members to thrive under a wide range of conditions, is remarkable. These free-sporing plants characterized by articulate stems bearing whorls of leaves at each node, exhibit diverse suite of adaptations that allow tolerance of disturbance, soil anoxia, high metals, and salinity, along with efficient nutrient uptake and nitrogen

*Scouring Rushes can be distinguished from other horsetails in the state by its large size, rough unbranched stems, and pointed cones.

fixation. All these characters account for the survival of living horsetails in diverse and harsh conditions making them the plants which provide insights into how their larger (tree-like) ancestors lived and how this ancient lineage has managed to survive in tropical regions.

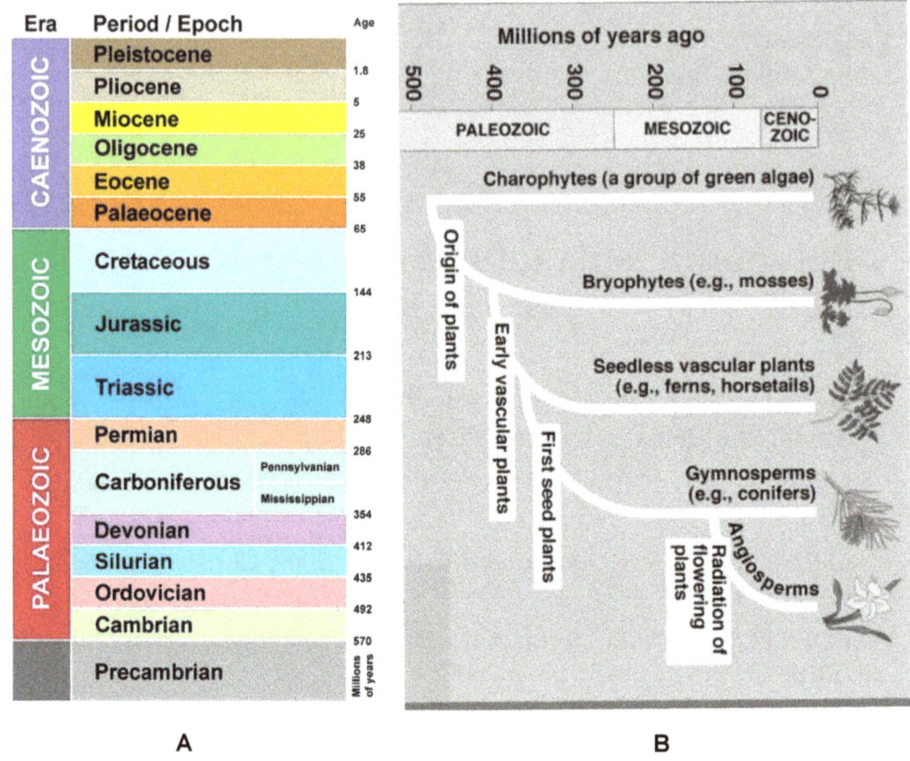

Figure 7.1A, B: *Equisetum* (The era and rise and fall of horsetails)

A - The Paleozoic era first of three eras within the Phanerozoic eon (the time of visible life) refers to the "time of ancient life" and spans the time period between 544 and 245 million years ago. B – Shows the records of Sphenophytes and plant groups.

SPHENOPSIDS OF THE PAST

The Sphenopsids are currently classified into two orders, the Sphenophyllales and the Equisetales (Stewart & Rothwell, 1993). The former consisting of a single genus *Sphenophyllum*, an herbaceous genus with whorls of wedge-shaped leaves on a jointed stem. The genus increased in abundance until the Upper Carboniferous, but disappeared by the end of the Permian. The Equisetales had three families Archaeocalamitaceae, Calamitaceae and Equisetacae. The Archaeocalamitaceae were arborescent sphenopsids which persisted from the Upper Devonian through the Lower Permian and were similar to the much more numerous Calamitaceae (Stewart & Rothwell, 1993). The Calamitaceae, with a single genus, *Calamites* attained heights of up to 30 m and diameters of up to 30 cm (Scagel et al., 1984). The family Equisetaceae consists of the living genus *Equisetum* besides other extinct herbaceous sphenopsids. Interestingly, the Calamitaceae closely resembled the Equisetaceae in having rhizomatous growth, and fused leaf sheaths at the nodes while the chief differences between the two families lie in cone morphology and in the lack of secondary (woody) growth in the Equisetaceae in contrast to the presence of secondary growth in the Calamitaceae (Stewart & Rothwell, 1993). The Carboniferous is considered the peak of pteridophyte diversity and abundance and it was also during this period that about 75% of the world's coal was formed. The great Carboniferous coal swamps were warm and humid and occupied the wet tropical low-lying areas (Pearson, 1995) and were dominated by giant arborescent Lycopods such as *Lepidodenderon* and *Lepidopholios* (Stewart & Rothwell, 1993). *Sphenophyllum* species were ground-cover plants which occurred in nearly all lowland habitats (Behrensmeyer et al., 1992). *Calamites* were hydrophytes, like *Equisetum*, and grew on loosely consolidated substrates such as sand bars, lake and stream margins, and other unstable moist substrates (Tiffney, 1985). Calamites were the only Carboniferous lowland arborescent plants with extensive vegetative propagation (Tiffney, 1985). The rhizomatous growth of Calamites, like that of modern *Equisetum*, allowed them to form extensive colonies on disturbed wetland areas. Then there occurred a worldwide change from hydric conditions to mesic conditions which were less favorable to sphenopsid growth. In addition, the inability of sphenopsids to grow in the increasingly dry sites reduced their ability to compete with the increasingly successful ferns, cycads, and conifers in these drier sites (Koske et al.,1985). These changes probably led to the extinction of Calamites during the Lower Permian and the extinction of the Sphenphyllales by the end of the Permian. These extinctions left Equisetales as the only representatives of the Sphenopsids (Stewart & Rothwell, 1993).

By the Mesozoic, all sphenopsids had acquired the same basic body plan as present day *Equisetum* (Behrensmeyer et al., 1992). The Equisetalean, *Schizoneura* an upright herbaceous genus, with stems up to 2 meters tall and 2 cm wide (Behrensmeyer et al., 1992), which first appeared in the Carboniferous, continued into the Jurassic (Stewart & Rothwell, 1993). *Schizoneura*'s large flat leaves were a distinctive feature, not commonly found in other members of Equisetales (Scagel et al., 1984). Another herbaceous sphenopsid which survived from the Carboniferous to the Lower Cretaceous was the genus *Phyllotheca* (Stewart & Rothwell, 1993). Both *Schizoneura* and *Phyllotheca* were at first confined to Gondwanaland (Gondwana was assembled by continental collisions in the Late Precambrian) where they dominated the broad swampy areas, but later due to migration, attained almost cosmopolitan distribution (Hallam, 1973). In addition, the genus *Neocalamites*, first appeared in the Upper Permian and survived until the Lower Jurassic.

Neocalamites widely distributed during the latter Triassic (Seward, 1959) resembled small *Calamites* in gross morphology (Stewart & Rothwell, 1993). It had stems 10 to 30 cm thick and about 10 m high (Behrensmeyer et al.1992). *Equisetites*, a genus which first appeared in the carboniferous, was the other major surviving genus of sphenopsids. The members were very similar to present day *Equisetum* and there is some controversy as to whether they may actually have been congeneric with present day *Equisetum*. If it was true then *Equisetum* can be said to have existed since the Paleozoic.

However, some Triassic and Jurassic *Equisetites* were significantly larger than present day *Equisetum*, reaching 8 to 14 cm in diameter (Stewart & Rothwell, 1993). Perhaps the largest *Equisetites* species, *E. arenaceus* that lived during the Upper Triassic period (Kelber & van Konijnenburg-van Cittert, 1998), had stems that averaged 25 cm in diameter and about 2.5-3.5 m in height. During the Jurassic, the large *Equisetites* were present in nearly all parts of the world. Stewart & Rothwell (1993) hypothesized that large *Equisetites* may have had secondary growth due to their size, but direct evidence for this is lacking, and it seems possible that the large *Equisetites* must have had lignified support tissues. While Spatz et al. (1998) did not find lignification in the supporting tissues of the *E. giganteum* stems they examined, Speck et al. (1998) found slight lignification in supporting tissues of *E. hyemale*.

The distribution and anatomy of Mesozoic sphenopsids was consistent with primary colonization of open or disturbed moist habitats. During the Triassic, the sphenopsids as a whole became less diverse and increasingly limited to herbaceous forms (Behrensmeyer et al., 1992) which was probably due to increasingly arid conditions during the Triassic. However, the surviving order *Equisetales* was widely distributed and diverse during the Mesozoic, but became smaller and less numerous from the Jurassic (Schaffner, 1930). By the beginning of the Coenozoic, relatively small species of *Equisetum* appeared (Stewart & Rothwell, 1992). This decrease in size and abundance during the Cretaceous was probably also related to the rapid rise of angiosperms to dominance and the resulting general decline in the prominence of pteridophytes and conifers (Schaffner, 1930). However, despite this decline, during the Quaternary, *Equisetum* species were found to be widely distributed in the temperate zone (Steward, 1959).

EQUISETACEAE: *FERN ALLIES OR FERNS?

The genus *Equisetum*, the common horsetail is the only living representative of the group variously referred to as 'Sphenophytes', Equisetopsida or Sphenopsida, ('Sphenophytes' - a term commonly used for extant horsetails and their fossil relatives) and which was once abundant and exhibited great diversity. Fossil literature usually traces the origin of the group in the Late Devonian (Stewart & Rothwell, 1993; Taylor & Taylor, 1993), culminating in considerable morphological diversity during the Carboniferous, when genera like *Calamites* with secondary growth formed trees dominating the forests of that time. *Equisetum* can be considered a 'living fossil' (Arnold, 1947) as it is the only surviving member of this putative clade (Box 7.1).

*The phylogenetic placement of *Equisetum* has remained challenging because of the controversy regarding its affiliation with extant ferns. *Equisetum* is an unresolved polytomy containing fossils that have been conventionally classified as ferns or fern-like plants (i.e., *Calamophyton* [Bonamo & Banks, 1966], *Eospermatopteris*, and *Pseudosporochnus* [Stein & Hueber, 1989]) along with representative fossil sphenophytes (i.e., *Calamites* and *Archaecalamites*) (Krings et al., 2018).

Recent research with molecular approach has however, indicated that Equisetaceae are not directly related to other Sphenopsida but have achieved their similarity to fern clade due to convergence and are an early-branching member of the clade (Pryer et al., 2001a; Lehtonen, 2011). Once placed along with club mosses under *fern allies, the present research done by various groups (Pryer et al., 2001) suggests that perhaps *Equisetum* should be classified within the true ferns – the Monilophytes (Fig. 7.2).

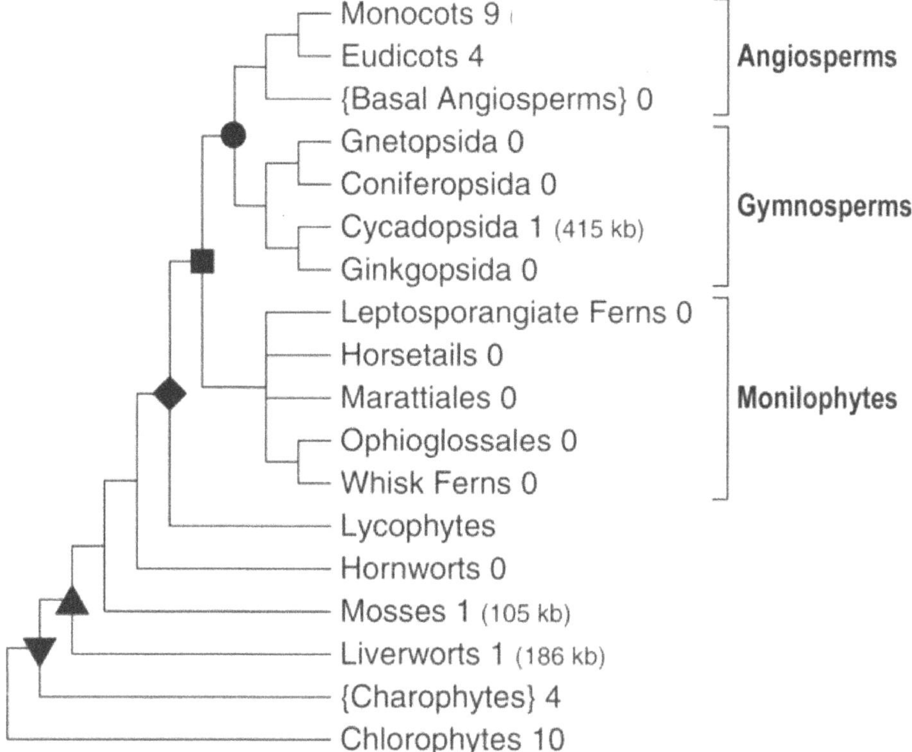

Figure 7.2: Current view of a simplified phylogeny of extant green plants sensu lato. The monophyletic clades of higher order are the spermatophytes (seed plants, circle), the euphyllophytes (square), the tracheophytes (vascular plants, rhomboid), the embryophytes (land plants, up triangle) and the streptophytes (down triangle) while the bryophytes (liverworts, mosses and hornworts) are paraphyletic.

* Fern allies are plants that have shared characteristics with ferns, but have distinct structural differences. For example, fern allies have smaller, undivided leaves. In addition, although fern allies reproduce by spores also, the sporangia are located in the axils or on the apex of sporophylls, as opposed to the underside of the leaf. These plants consist of fork-ferns, tassel ferns, whisk ferns, horsetails, club mosses, and quillworts. There are 1,600 species of fern allies.
It may be noted that based upon recent molecular data the ferns are now also considered to include the horsetails (*Equisetum*) & whisk ferns (*Psilotum*) (February 7, 2017** - The higher order classification of ferns has been reorganized to correspond with: *PPG I (2016), A* community-derived classification for extant lycophytes and ferns. Jnl of Sytematics Evolution, 54: 563–603. doi:10.1111/jse.12229. Based on this the classification is: Polypodiopsida - Equisetidae (horsetails) - o09i8u809o09)

BOX 7.1

Equisetum spp commonly known as horsetails, are the living descendants (living fossils) of giant prehistoric trees that once dominated the Earth. These ancient giant trees remain today in the form of coal. The living species of *Equisetum* offer biologists a peek into the history and evolution of vascular land plants as it is the single extant genus of a class of anciently evolved land plants. Fossils of the ancient giants date back to the upper Devonian, and in the late Devonian and Carboniferous periods, when the world's forests were dominated by the *Calamites* — huge tree-like horsetails. The giant arborescent *Calamites* that reached 30 m in height, formed an understory in Carboniferous coal swamps, and became extinct in the lower Permian (Stewart & Rothwell, 1993).The *Calamite* forests were so rich that they eventually formed the world's large coal deposits. As the climate grew drastically drier and colder in the Permian, concomitant with a change in atmospheric oxygen content, the giant *Calamites* eventually died out, leaving only the herbaceous genus *Equisetum* as the representative in modern times. *Equisetum* species are almost identical to the ancient *Calamites* in morphology, earning *Equisetum* the label of 'living fossils'. Various equisetal fossils are known, but the earliest lineages that can be assigned to *Equisetum* appear in the Triassic and have been described under the generic name *Equisetites*. One of the earliest species placed in this genus was the upper Triassic *E. munsteri* (Sternberg, 1853). Horsetails are found in various habitats around the world, though they usually prefer wet, swampy areas. Although 'giant' horsetails (e.g. one Mexican species) are reported to reach heights of 8 meters, most are much smaller, usually around 1 meter shorter than their ancestors.

EQUISETUM

Equisetum, the genus with 15 living species comprises plants commonly known as horsetails. The name *Equisetum* is derived from the Latin word – 'equis', meaning horse and seta, meaning bristle, which refers to the coarse black roots of *E. fluviatile* resembling a horse's tail. The horsetails range in size from small sized *E. scripoides* (stems averaging 12.9 cm tall and 0.5 - 1.0 mm diameter) to the giant horsetails, *E. giganteum* and *E. myriochaetum,* which reach heights of 8 or more meters (Hauke, 1963) and stem diameters of about 4 cm. *Equisetum* species are vascular plants which reproduce sexually by means of spores borne on cones also referred to as strobili. Hence, together with the other spore-bearing vascular plants, the Lycophytes (club mosses), Psilophytes (whisk ferns) and Pterophytes (true ferns), *Equisetum* species have been grouped as pteridophytes.

TAXONOMIC CONSIDERATIONS

The present day genus *Equisetum* is divided into two distinct subgenera: subgenus *Equisetum*, with eight species and subgenus *Hippochaete*, with seven species (BOX 7.2). There are several primary differences between the two subgenera. Species in subgenus *Equisetum* have stomata that are present in the epidermal surface, whereas in subgenus *Hippochaete*, members possess stomata that are sunken below the epidermal surface. The stems of the subgenus *Equisetum* are short-lived, relatively soft, and tend to be regularly branched, whereas the stems of the subgenus *Hippochaete*, tend to be long-lived, hard, fibrous, and unbranched or irregularly branched, though exceptions exist. In addition, four of the species of the subgenus *Equisetum* demonstrate stem dimorphism between non-photosynthetic, unbranched stems that bear storobili (coniferous stems) and photosynthetic, branched, vegetative stems. No such stem dimorphism occurs in the subgenus *Hippochaete* (Hauke, 1963). Although the chromosome number (n=108) is the same for all *Equisetum* species, the subgenus *Hippochaete* has larger chromosomes than those of subgenus *Equisetum* (Hauke, 1978). The subgenus *Hippochaete* includes species often called "scouring rushes" (although also known generally as horsetails) due to their rough, silica impregnated epidermis. The rough silicaceous stems of plants were used by American pioneer settlers for scouring dirty cookware and polishing wood (Scagel et al., 1984). The seven species in this group are *E. giganteum*, *E. myriochaetum*, *E. ramosissimum*, *E. laevigatum*, *E. hyemale*, *E. variegatum* and *E. scirpoides* (including the two largest *Equisetum* species, *E. giganteum* and *E. myriochaetum*). With the exception of *E. laevigatum*, and some varieties of *E. ramosissimum*, all of the species in this subgenus have evergreen stems (Hauke, 1963).

Hippochaete is widespread with species distributed over large areas of every continent, except for Australia and New Zealand. The Old World species *E. ramosissimum*, which ranges from 60° North latitude to 30° South latitude, has the widest latitudinal range of any *Equisetum* species (Schaffner, 1930). The subgenus *Hippochaete* ranges as far north as Ellsmere Island (greater than 80° North latitude) and as far south as Argentina (approximately 40° South latitude) (Hauke, 1963).

The subgenus *Equisetum* contains the species commonly known as "horsetails." The eight species of this group are *E. arvense*, *E. pratense*, *E. sylvaticum*, *E. fluviatile*, *E. palustre*, *E. bogotense*, *E. diffusum*, and *E. telmateia*. The species in this group tend to be regularly branched and hence can resemble bushy horsetails. Certain members of this subgenus are found from 80° North latitude to 40° South latitude. Only one species of this subgenus, the diminutive *E. bogotense* of Central and South America, has a range that extends to the Southern Hemisphere. The other seven species of this group are found in the Northern Hemisphere (Hauke, 1963). Most species of subgenus *Equisetum* are temperate, with a few extending their ranges into the subtropics and only *E. bogotense* ranging into the tropics. The aerial stems of all of these species, except for *E. bogotense* and *E. diffusum*, (the two most southerly species), are annual (Hauke, 1978).

BOX 7.2

Christenhusz et al., (2019) based on phylogenetic studies recognize 18 species in three subgenera: subgenus *Paramochaete* (one species: *E. bogotense*); subgenus *Equisetum* (eight species: *E. arvense, E. braunii, E. diffusum, E. fluviatile, E. palustre, E. pratense, E. sylvaticum, E. telmateia*); subgenus *Hippochaete* (eight species: *E. giganteum, E. hyemale, E. laevigatum, E. myriochaetum, E. praealtum, E. ramosissimum, E. scirpoides, E. variegatum, E. xylochaetum*). In addition, two varieties are accepted in *E. ramosissimum*: *E. ramossisimum* var. *huegelii* and var. *ramossisimum*. **Source:** Phylogenetics, classification and typification of extant horsetails (*Equisetum*, Equisetaceae) Maarten J. M. Christenhusz*, Lois Bangiolo, Mark W. Chase, Michael F. Fay, Chad Husby, Marika Witkus and Juan Viruel, 2019. Botanical Journal of Linnean Society, 189:311-352.

HABIT

Equisetum species are herbaceous, evergreen, spreading, reed-like perennials with cylindrical, hollow, photosynthetic stems that bear tiny appressed, residual leaves. The stems are jointed with node-internode construction, usually unbranched, and with longitudinal ridges.

Most of the species are less than one metre (three feet) tall with reports of specimens of *E. giganteum*, from the American tropics, attaining a height of about 10 metres and a stem with diameter of about 4 centimetres (1.6 inches). For this kind of habit, support is apparently provided by their habit of growing in dense stands and by the vegetation that surrounds them in their natural environment.

DISTRIBUTION

Horsetails grow in full sun to partly shaded habitats and tolerate total immersion of their roots in water. It spreads aggressively in the shallow edges of water bodies along streams, ditches, and canals; some species, however, have become adapted to drier and sunnier conditions. Present day *Equisetum* species although are naturally distributed throughout the world, they are notably absent from Australia and New Zealand (Scagel et al.,1984) and from the islands of the central Pacific, Indian, and South Atlantic islands (Schaffner, 1930).

E. arvense is cosmopolitan in distribution, throughout Europe and Asia, south to Turkey, Iran, the Himalayas, and across China (except the south eastern part), Korea and Japan. It is also found throughout Canada and the USA as far south as Georgia, Alabama, Arkansas, Texas, Arizona, New Mexico and California (Hultén & Fries, 1986). According to Holm et al. (1991), *E. arvense* is a principal weed in Belgium, Canada, England, Finland, Germany, Japan, New Zealand, the former Soviet Union, the USA, the former Yugoslavia and is a common weed in Alaska, Argentina, Brazil, the former Czechoslovakia, France, India, Iran, Madagascar, Mauritius, Netherlands, Poland, Romania, Spain and Sweden. In Chile, China, Iceland, Italy, Korea and Turkey it is only reported as present. It has adapted to a range of habitats outside its preference for damp, acidic soils. Once the underground rhizomes become established, the plants can be quite difficult to eradicate because they are resistant to many herbicides and the plants become invasive.

**Equisetum* species are avoided by most insects, probably because of the abrasive silicates associated with the outer epidermal cell walls (Kaufman et al., 1971) and possible toxic compounds within the cell sap. However, some insects have formed obligate host associations with horsetails, while other insects are occasional feeders.

SPOROPHYTE

Morphology

The morphology of *Equisetum* is highly unique among living vascular plants. The plants consist of upright aerial stems and a very extensive underground rhizome system, both of which are jointed through nodes and internodes, latter being ribbed (Fig. 7.3A). The upright aerial stems exhibit a monopodial branching, having one main axis of growth, a pattern which is also found in most gymnosperms and angiosperms (Scagel, et al., 1984). The microphyllous leaves are arranged in true whorls (Rustishauser, 1999) and the leaves of each whorl are fused together towards the base to form a cylindrical sheath around each node (Hauke, 1993). The whorls of lateral branches at the nodes of the aerial stems are also seen in some species (Fig. 7.3B, **PLATE 4A**). Unlike many other vascular plants (such as gymnosperms, angiosperms and some ferns) which also produce branches in the axils of leaves, the leaves of *Equisetum* are unique in that they alternate with branches at each node (Scagel et al., 1984). Like other vascular plants, *Equisetum* possesses an apical meristem from which arise branches and leaves. However, most of the stem lengthens by the activity of intercalary meristems present above each node.

The aerial stems, but not the rhizomes, of some species die back seasonally, acting as vegetative means of reproduction while rhizomes continue to grow. The rhizomes are constructed on the same general morphology as the upright stems, although they bear adventitious roots at their joints in addition to leaf sheaths and branches. The plants range in size from the 9 m high tropical species *E. myriochaetum* to the 4-5 cm tall temperate species, *E. scirpoides*.

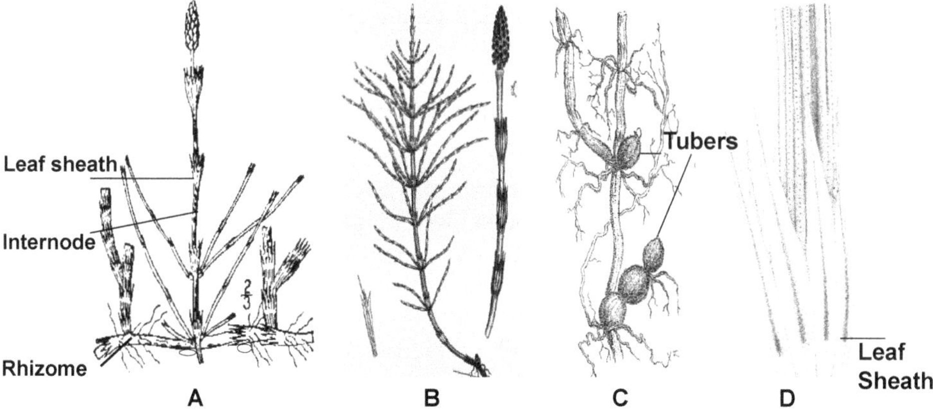

Figure 7.3A-D: *Equisetum* – (Morphology)

A - The plants consist of upright aerial stems and a very extensive underground rhizome system, both of which are jointed through nodes and internodes, latter being ribbed. The rhizome bears tubers while the terminal strobili are borne on fertile aerial shoots. B – Portion magnified to show lateral branches constructed on the same plan as the main shoots. C- Portion of the rhizome magnified to show tubers. D- Leaf sheath formed by the leaves at a node by their fusion at the base. The free ends appear teeth-like.

The sporophyte consists of an underground rhizome from which arise nodal buds giving rise to upright aerial shoots. The subterranean rhizomatous system which bears roots, tubers (Fig. 7.3C) and branches is extensive and responsible for the spread of the genus.

- The plants appear jointed or articulated in appearance because the stem is constructed on node-internode pattern where nodes are typically separated by relatively long internodes.
- The internode has a series of ridges and valleys or furrows which continue up into the nodal sheath. The successive internodes show alternating ridges and furrows (a ribbed appearance). Each rib corresponds to the main vascular bundle, the upper part of which is a leaf trace.
- The internode may continue to grow for some time through the activity of a meristem located at its base.
- At the nodes are present reduced leaves that remain united at least for a part of their length, forming a sheath around the stem (Fig. 7.3D). The number of leaves at a node corresponds to the number of ribs on the internode below.
- Lateral branches when evident, are attached at the nodes and alternate with the leaves. Branches which may arise regularly or irregularly and at times as a response to injury, are constructed on the same plan as the main stem.
- Fertile organs consist of peltate sporangiophores that are also borne in whorls around the central axis, and form terminal bractless compact strobili on main aerial stems.
- The sporangia are borne on the undersurface of peltate disc of the sporangiophores.

**In some species, the nature of branches is characteristic with reproductive and vegetative branches differentiated in the sporophyte. The fertile branches found above the ground, bear strobili at their apical ends (terminal). These branches arise first and shed spores from their sporangia even before the appearance of the sterile branches. In *E. arvense* very early in spring, ephemeral and deciduous, brownish fertile stems that are unbranched and devoid of chlorophyll, arise from rhizomes breaking through the soil. They later bear terminal strobili rounded at the tips and they die soon after the spores are shed. Later in spring, the same rhizome produces green, infertile, vegetative branching systems. They have whorls of ascending branchlets along at least the upper two-thirds of their stems. The stems and branchlets of these shoots are slender, glabrous, and green. These branchlets are unbranched. An infertile shoot may rarely produce a small infertile strobilius at its apex that aborts prematurely.

E. palustre, E. cylvaticum and *E. pratense* exhibit great specialization and segregation of vegetative and reproductive systems: (a) the sterile branches which are deep green and much branched, (b) the fertile shoots which are brown, short lived, unbranched and devoid of chlorophyll, (c) the intermediate type of branches which are fertile, devoid of chlorophyll and unbranched at first, but later, the strobili are thrown off and the branch becomes green, branched and persistent. This is an example of division of labor in *Equisetum* plant (BOX 7.3).

Box 7.3

On the basis of external morphology alone, the evolutionary tendencies in the genus have been summarized:

The more primitive species have perennial, green shoots that are highly branched and bear green, pointed strobili, terminating several branches of the same shoot. The most advanced species have the annual, branched aerial vegetative shoots that are green and the short-lived unbranched fertile shoots which lack chlorophyll but bear reproductive structures. An example of *Equisetum* with two types of aerial branches is *Equisetum arvense*, a species that is often found along railroad right-of-ways. Such morphological plasticity is expressed through evolution amongst the members of the group. In extant *Equisetum*, the internodes that make up the shoots are exclusively vegetative, whereas sporangiophores are found only on specialized shoot tips with determinate growth (strobili). In contrast, in the extinct genus *Cruciaetheca* and in species of the extinct genera *Equisetinostachys, Tschernovia* and *Paracalamitina* some internodes bear whorls of sporangiophores along part or all of their length; such internodes are usually aggregated in fertile regions along stems. In yet another group of fossils, such as *Peltotheca* and in species of the extinct genera *Pothocites* and *Archaeocalamites* vegetative internodes are replaced in the terminal region of specialized fertile shoots by internodes covered in sporangiophores. The morphological diversity encompassed by these fossils has raised fundamental questions about the origin, homology, and evolution of equisetacean reproductive structures, and the answers need to be worked out.

Anatomy

Stem - Internode

The outline of the section of internode is circular with wavy outline marked by ridges and furrows which correspond to the ribbed construction of the internode

(Fig. 7.4A). Outermost layer is epidermis which remains covered with silica layer. Stomata are generally present in the furrows and remain sunken below the surface. The silica is deposited in the wall of guard cells in transverse radial bands. The sclerenchyma remains massed under grooves of the stem and provides mechanical support to the stem. The cortex is generally differentiated into outer and inner cortex where outer is chlorenchymatous and inner is parenchymatous. A ring of large canals known as vallecular canals is present in the cortex opposite each furrow and provides an aerating system as the canals are filled up with air. The stele inner to cortex is ectophloic siphonostele which remains surrounded by endodermis, inner to which is pericycle. The vascular bundles are arranged in a ring and are placed opposite to the ridges alternating with the vallecular canals of the cortex (**PLATE 4B-D**). Associated with the vascular system is an endodermis around each bundle or there may be a common endodermis around a ring of vascular bundles (Fig. 7.4B). In certain species as *E. hyemale,* there is double common endodermis - one outside the ring of vascular bundles and one inside. Vascular bundles are conjoint, collateral and closed. An additional ring of smaller canals is present deeper within the stem. These mark the position of the first-formed xylem, or protoxylem (which is endarch), that becomes disorganized as the stem elongates and matures. The resulting canals which develop in each bundle are called carinal canals (alternate with the vallecular canals). The carinal canals are filled up with water.

The remaining protoxylem elements are composed of few tracheids. The metaxylem elements are found in two groups, arranged on the margin of the carinal canal towards outside (Fig. 7.4C). The protoxylem lies in between the two groups of metaxylem. The metaxylem is generally composed of reticulate, scalariform or pitted tracheids while spiral and annular tracheids are also occasionally found. The phloem is composed of sieve tubes and phloem parenchyma. The sieve plates may also be seen. The companion cells and secondary growth is altogether absent. The central region of the internode in young stem is occupied by a parenchymatous pith but at maturity it is represented by a central canal.

*The nodal region is reported to have a little variation when compared to the internode. The carinal canals are absent as the protoxylem elements are intact in the nodal region. At the node, stem is not hollow but the pith forms a diaphragm separating the two successive internodes (**PLATE 4E**).

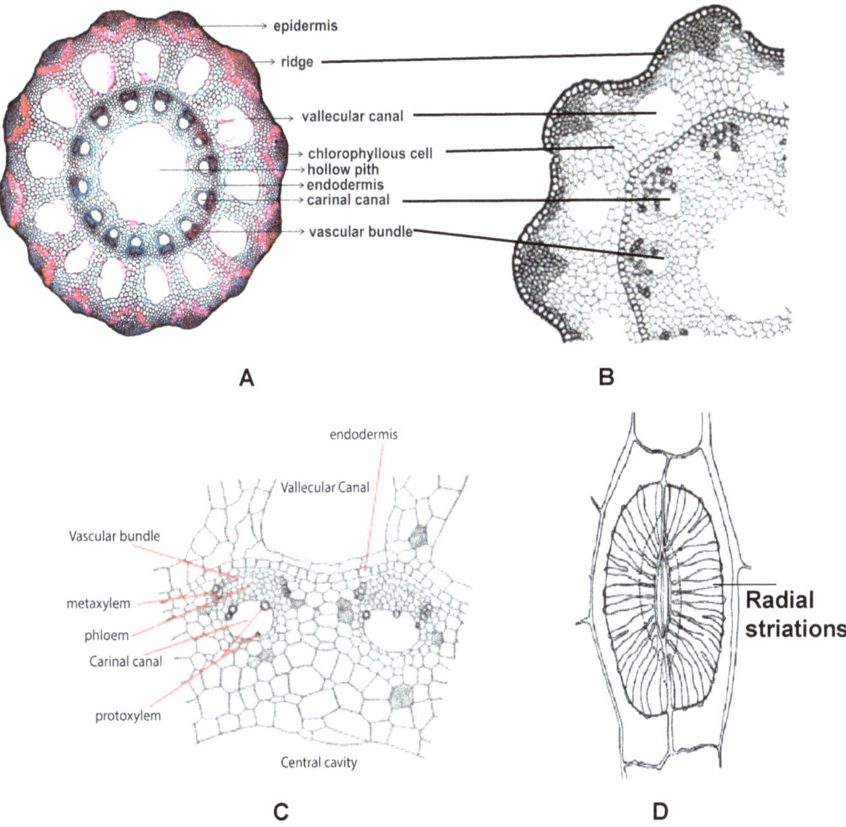

Figure 7.4A-D: *Equisetum* – (Anatomy)

T.S. Stem through the internode. A - Outline diagram. B – Cellular details of the same to show ridges and furrows, vallecular and carinal canals and the central cavity. In C - the vascular bundles are magnified to show proto- and meta xylem and phloem. D- To show stomata in the stem. Note the radial striations in the guard cells that represent silica deposition.

Salient features of Stem (internode)

- Epidermis single layered with stomata present in furrows to check transpiration.
- Silica deposition over epidermis and over guard cells (Fig. 7.4D) makes stem resistant to stress.
- Sclerenchyma below the ridges gives mechanical support and chlorenchyma which forms outer cortex compensates for the non-photosynthetic leaves.
- Inner cortex is parenchymatous and shows air filled vallecular canals below the furrows.

- Endodermis and pericycle seen; the former may be double in some species.
- Vasculature is eusteleic, with a ring of collateral and closed vascular bundles present below the ridges.
- Vascular bundles at maturity show carinal canals below the ridges; canals are formed by disintegration of protoxylem elements and xylem is therefore, poorly developed.
- Pith is large and is generally replaced by a cavity at maturity.

Adaptive features of stem

- Stems of horsetails are "anatomically unique among plants" (Niklas, 1997).
- The stem of *Equisetum* contains very little vascular tissue and is entirely primary, even in the tropical forms that often reach 20 feet in height.
- In most species the central portion of the stem (pith) becomes hollow in the internodes.
- Presence of reduced xylem, and the well-developed aerating system formed by vallecular and the central pith cavity is a hydrophytic character.
- The stem exhibits certain xerophytic characters as well and are adapted to reduce water loss. The thick cuticle, thick epidermis, stomata in furrows and silica depositions and cover cells and carinal canals filled with water, help plants check transpiration and deal with water economy.
- The carinal canals in *E. ramosissimum* have a distinctive lining containing pectic homogalacturonan, cellulose, xyloglucan and extensin. These canals might function as water-conducting channels which would be especially important during the elongation of the internodes when protoxylem is disrupted and the metaxylem is not yet differentiated. How the molecularly distinct lining relates to the proposed water-conducting function of the carinal canals requires further study (Leroux et al., 2011).
- The silica deposits in *Equisetum* sp. indicate a physiological mechanism of stress tolerance as silicon uptake has been demonstrated to ameliorate salt, heat and heavy metal stress in other crops.

Morphometric and mechanical characteristics

The stem internodes (studies done on *E. hyemale* L.) possess mechanical characteristics that influence the capability of the stem to vibrate, enabling spore liberation at minimum stress. *Equisetum sp.* present special biomechanical features besides a simple body structure. It consists of an unbranched stem which attains the height of 150 cm, with basal width of 4–6 mm and characteristically ridged internodes and nodes (Niklas, 1989b; Speck et al., 1998). Anatomically it is hollow due to the formation of a large, central pith cavity and the vallecular and carinal canals. The thick walled sclerenchyma present below the epidermis acts as a mechanical support. While the central canal qualifies horsetails as a hollow tube structure, the transverse nodal septa resist stem to bending and twisting (Niklas, 1997a). Unlike grasses which also have hollow stems, *E. hyemale* is able to resist mechanical collapse because of changes to cellular water status of the cells during the vegetative period. They also have the ability to resist sub-freezing temperatures, due to extracellular freezing phenomena (Niklas,1989a), whereby shoots remain erect after thawing processes. The latter feature is almost independent of hydration state and relates to the ability of shoots to resist local buckling or sudden bending. This is because of the presence of a double layer of endodermis (Spatz et al., 1998; Speck et al., 1998). There is variation in length of horsetail stem internodes, as well as their width and strength properties as one moves from the base to the plant apex. Internodes at the base and in the apical region of the stem are shorter and thinner than those in the middle part of the plant.

Leaf

- The anatomy of the leaf is simple as the leaves are uninerved.
- The vascular bundle of the leaf (that forms a sheath) is simple and collateral and each bundle of the sheath is surrounded by a separate endodermis.
- The outer tissues of the leaf sheath are in form of sclerenchymatous bands that pass up the leaf ridges and alternate with the strips of chlorophyllous tissue associated with stomata.

Root

- The adventitious roots are borne at the nodes of the rhizome and aerial shoot.
- The outermost layer is piliferous layer, which bears unicellular hairs.
- It is followed by a multilayered cortex. In the small roots, the cortex is divided into two zones - the outer zone composed of three to four layered lignified exodermis and the inner zone of thin walled parenchyma with well-defined intercellular spaces (**PLATE 4F**).
- The endodermis is two celled in thickness where the inner layer may act as pericycle which otherwise is generally absent. The lateral roots originate from the inner layer of the endodermis. Small intercellular spaces are found between the two layers of the endodermis. The outer layer of endodermis has casparian thickenings (Fig. 7.5A).
- The stele of the root varies in the nature from species to species. It is triarch to hexarch with a large central metaxylem element surrounded by three to six narrow points of protoxylem represented by single tracheid (Fig. 7.5B). The tracheids of xylem elements are spirally thickened. The angles between the protoxylem points are completely filled with phloem composed of phloem parenchyma and sieve tubes.

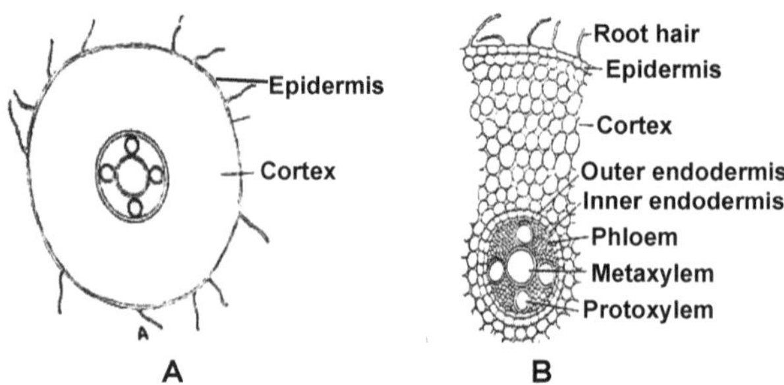

Figure 7.5A,B. *Equisetum* – (Anatomy)

A. T.S. Root - Outline diagram. B - Same showing the central metaxylem, and four protoxylem elements around it. Also seen is the outermost piliferous layer, an undifferentiated cortex and double endodermis.

Reproduction

The reproductive structures of *Equisetum* species are blunt strobili (or cones) growing on the tips of fertile yet hollow stems (**PLATE 4G**). They consist of a central axis to which aggregates of sporangiophores are attached. The sporangiophores bear sporangia, each containing numerous spores which are green, small and uniform in size, and have elaters for dispersal. The plant is homosporous.

Structure of Strobilus and Sporangiophore

- The strobili are terminal, narrowly ellipsoid or oblong-ellipsoid with whorled and stalked umbrella-like or peltate sporangiophores (**PLATE 4H**).
- The lowermost sporangiophores may be sterile.
- In a section, pad-like rhomboid sporangiophores (also referred to as sporophylls) are seen arranged in whorls along a central axis (Fig. 7.6A,B).
- Usually, each such whorl is composed of twenty or so sporangiophores.
- In many cases, immediately below the sporangiophores, the central axis of the strobilus bears a small ring-like outgrowth known as the annulus.

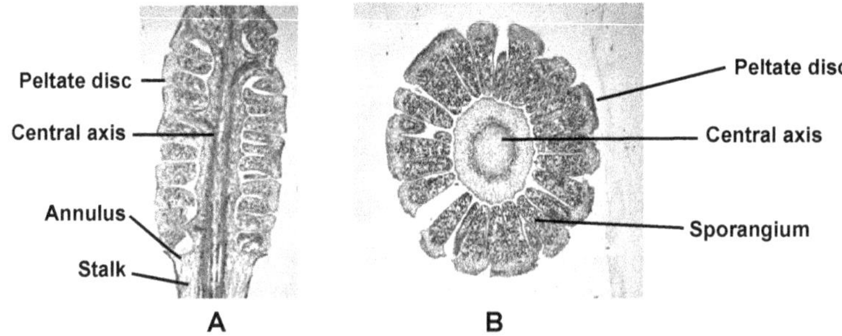

Figure 7.6A,B: *Equisetum* (Reproductive)

A - Longitudinal section and B- Transverse section of the strobilus to show a central axis with sporangiophores arranged around it. Also seen are the sporangia on the under surface of the peltate discs of the sporangiophores.

Sporangiophore

- All the sporangiophores exhibit symmetry in their arrangement around the central axis.
- Each sporangiophore is composed of a single slender stalk attached to the central axis.
- The terminal end of the sporangiophore is expanded into a flattened peltate disc situated at right angles to the stalk (**PLATE 4I,J**).
- The peltate disc is usually hexagonal in outline in its surface view.
- This hexagonal appearance of the discs is attained due to the mutual pressure of the discs of the crowded sporangiophores situated on the strobilus.
- Each peltate disc bears a ring of 5 to 10 isolated inward-pointing sporangia on its underside (Fig. 7.7A).
- Each sporangium is elongated and sac-like. These sporangia may be described as being recurved toward the stem (Fig. 7.7B).

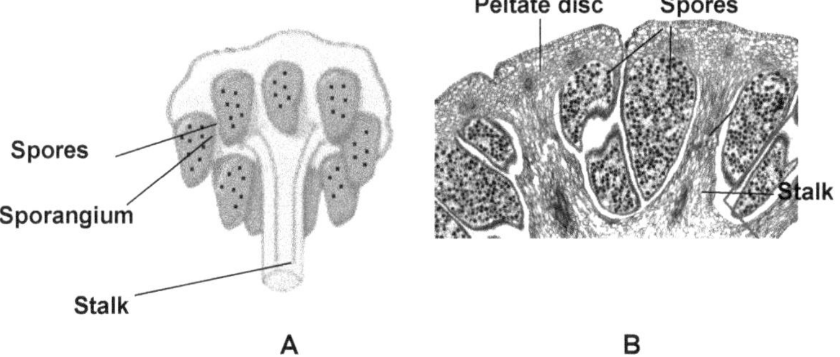

Figure 7.7A,B: *Equisetum* showing a sporangiophore with a peltate disc and a stalk. In B- spores are seen inside the sac-like sporangia. The sporangia have a distinct wall and tapetum which enclose spores.

Sporangium

The development of the sporangia of *Equisetum* is eusporangiate type – i.e., it is derived from a group of sporangial initials. However, Bower believed the entire sporogenous tissue in a sporangium develops from a single superficial cell of the young sporangiophore.

- Each mature sporangium is elongated, sac-like and rounded at its apex (**PLATE 4K,L**).
- On maturity, it consists of at least two layers of which the inner is tapetum.
- The sporangium contains within it the mature spores which are all alike.
- *Equisetum* is therefore, homosporous.

Dehiscence of Sporangium

- On maturation of the spores the sporangium, bursts by a longitudinal slit.

- Prior to the opening of the sporangia, the central axis of the strobilus elongates separating the sporangiophores quite apart from each other (**PLATE 4M**).
- As the sporangium dries up, the walls of the cells of the jacket layer which are spirally thickened, shrink and lose water and the longitudinal slit develops on the sporangium for the release of the spores.
- With further drying up, the thickened bands present in the outer wall layer shrink further and the sporangium ruptures.

Spores

- The spores are spherical, uninucleate and green, and contain numerous chloroplasts.
- The spore wall is differentiated into four concentric layers - On the extreme outside there is the spirally cleft wall, perispore or epispore which constitutes the curiously hygroscopic structure - the elater.
- Within this layer is a membrane which is usually spoken of as the "middle layer"; inner to this lies the "exospore," whilst the innermost layer of all is the endospore.
- The spores have four elaters, which are flexible ribbon-like appendages with expanded spoon-like tips and are hygroscopic responding to change in humidity. They are initially wrapped around the main spore body and that extend or uncoil upon drying or coil or fold back in humid air **(PLATE 4N)**.
- As the spores dry out, the elaters snap open violently, separating the spores to be partially dispersed (Fig. 7.8A,B).
- Although these four elaters are separate from one another, they remain attached at a common point on the spore wall.

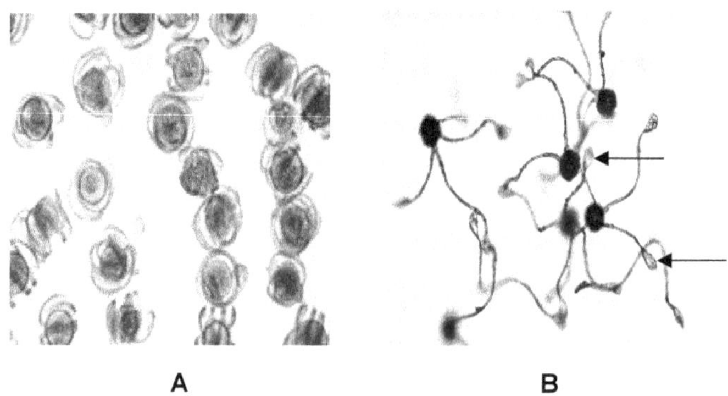

Figure 7.8A,B: *Equisetum* WM (Spores and elaters)
A – Wet preparation with elaters coiled around spores and B- Dry preparation with elaters extended to show spoon shaped ends (arrows).

Functions of Elaters

- The function of elaters is not very certain, but it is assumed their expansion may help in the dehiscence of sporangium.
- It is also thought that their hygroscopic movements may assist in the dispersal of the spores. According to some workers due to the presence of elaters, the spores are entangled together and thus dispersed in groups so the prothalli are found in close vicinity.
- They may also act as parachute helping spores to float in air during dispersal.
- According to Goebel thc elaters help in fastening the spore to the substratum (see Box 7.4).

Box 7.4

The movement of spores with the help of hygroscopic elaters is described as the 'walks' and 'jumps' of *Equisetum* spores (Marmottant P, Ponomarenko A, Bienaime´ D. 2013. The walk and jump of *Equisetum* spores. Proc R Soc B 280: 20131465. http://dx.doi.org/10.1098/rspb.2013.1465). These are novel types of spore locomotion mechanisms compared to the ones of other spores such as those of many fungi. Walks are driven by humidity cycles, where each cycle induces a small step in a random direction. The dispersal range from the walk is limited, but the walk provides key steps to either exit the sporangium or to reorient and refold. First, with humidity variations or rain, groups of spores make large steps, which helps their fall from the sporangium. It is an additional contribution to the spilling out of spores owing to the volume increase of clusters when spores unfold. Second, the random steps reorient the spores, and change their contact points with the ground. This results in different entanglements at each humidity cycle, and increases the probability to block the elaters and store elastic energy. Third, the reorientation also helps to obtain jumps in random directions, maximizing the explored area.

Jumps occur when the spores suddenly thrust themselves after being tightly folded. They result in a very efficient dispersal, as spores jumping from the ground can catch the wind again, whereas non-jumping spores on the ground, make gliding movement. When drying from a fully hydrated state, spores can suddenly leave the ground at speeds of approximately 1 m s^{-1} and reach elevations of up to one centimetre. This height is very large compared with the spore size. Jumps enhance the dispersion of spores and provide a way to overcome adhesion to the plant or to neighbouring spores. The jump allows the spore to exit the ground where the air velocity is reduced, and to enter the higher velocity wind current. These spores are then more likely to travel than are immobile spores. Since the jumps occur upon drying, they provide an escape from dry locations and thus increase the chances of finding a more humid environment. The understanding of these movements, which are solely driven by humidity variations, conveys biomimetic inspiration for a new class of self-propelled objects.

GAMETOPHYTE

Germination of Spore and Development of Prothallus (Gametophyte)

- The spores survive only for a few days after their liberation from the sporangium, later germinating in the clayey soil and mud along the banks of the rivers and streams.
- Prior to germination, the spore increases to some extent in its size and after a few divisions gives rise to one to several celled thick cushion-like tissue known as gametophyte or prothallus. The prothallus bears numerous rhizoids on its lower surface.
- The green lobes or branched structures are found on the upper region.
- Further development results in the formation of a prothallus with three distinct regions: i) The upper erect, green, photosynthetic portion in the form of spongy, irregularly-shaped lobes, ii) The middle basal prostrate region of light-yellow colour, iii) The lowermost region of colourless cells that gives off rhizoids.
- In certain species the gametophytes are quite big in size, e.g., in *E. debile* the gametophyte is about 3 centimeters in diameter (Kashyap, 1914), and two to three millimeters in thickness. In majority of the cases the gametophytes are less than a centimeter in size.
- The gametophytes are long lived, surviving in some cases for about two years.
- The prothallus is generally attacked by a fungus in the upper cells of the lobes.
- Internally prothallus is differentiated into two zones: (i) lower compact rounded parenchymatous portion forming the disc, and (ii) an upper spongy portion.
- The disc is composed of non-chlorophyllous large cells which are compactly arranged and are full of starch grains. The disc has outer marginal meristematic rim which increases the diameter of disc and forms new erect lobes as well as rhizoids.
- When the growth of the entire margin is uniform, the prothallus becomes quite smooth and circular, while on the other hand when growth is more active on one side or the other, the irregularly lobed prothallus is resulted.
- The spongy upper portion is composed of densely crowded green vertical lobes which completely cover the disc below. The lobes are irregular, plate-like expansions of chlorophyllous tissue several cells thick at the base but higher up becoming thinner and thinner, the ultimate part being only one cell in thickness (Fig. 7.9, **PLATE 4O**).

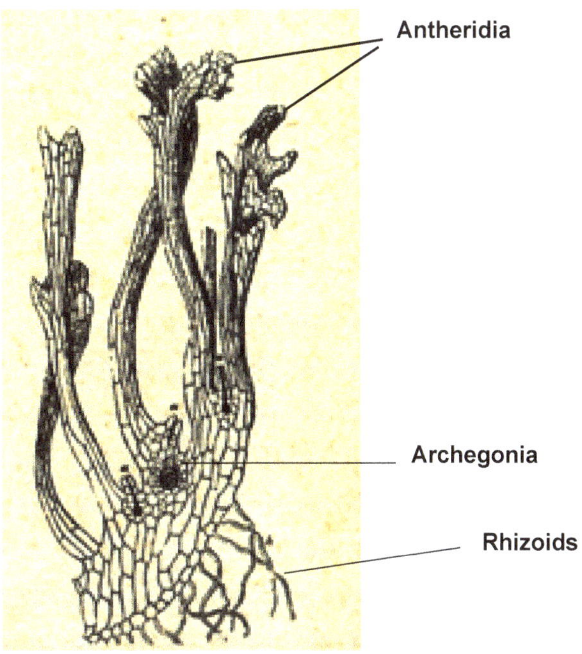

Figure 7.9: *Equisetum* (Prothallus)

V.S. Prothallus shows a parenchymatous disc from which arise rhizoids and the upper spongy lobed chlorophyllous region that bears sex organs.

Nature of prothalli

There are two types of prothalli - the small (male) prothalli bear only antheridia (Fig. 7.10A) whereas the large hermaphrodite prothalli bear both the antheridia and archegonia (Fig. 7.10B). Whether *Equisetum* is monoecious or dioecious is a matter of controversy and several views have been expressed. According to Kashyap's research, in *E. debile* thickly sown spores produce small prothalli which bear only one type of sex organ. On the other hand, if the spores are sown quite separate from each other, the prothalli become large in size and bear both male and female sex organs, i.e., antheridia and archegonia. In such cases, however, the archegonia develop first followed by antheridia (1917). According to Walker (1921), in *Equisetum arvense*, under unfavorable conditions the prothalli remain small in size and only antheridia develop on them. These antheridia reach maturity and release viable male gametes (Fig. 7.10C). In favorable conditions, half of the prothalli are developed as males and the remaining half as females. The normal prothalli are monoecious, which develop archegonia first and the antheridia later. He also stated that the prothalli developing in crowded conditions remain small and are dioecious. *Equisetum arvense* is mostly accepted as being monoecious. The gametophytes of *Equisetum* are potentially bisexual, but sequential development of the antheridia and archegonia generally ensures cross-fertilization.

Not only the gametophyte, but even spores vary in size and sex organs formed. According to Joyet Lavergne (1931), morphologically the spores are alike (homosporous), but physiologically they are different, and in the same lot some of the spores are male and some are female, implicating internal heterospory.

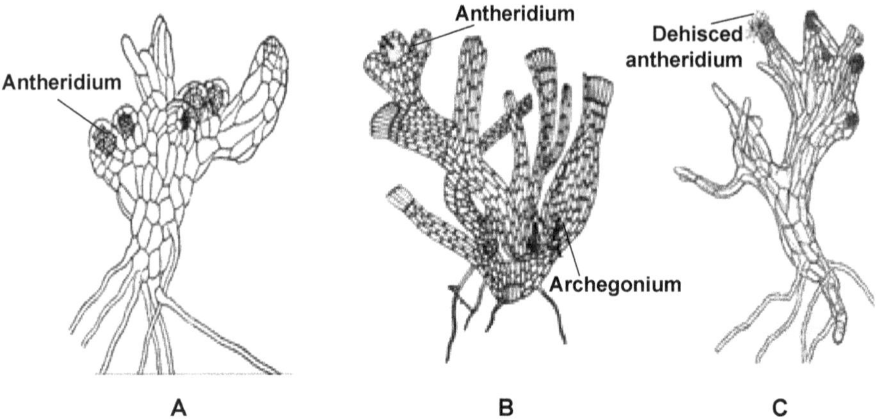

Figure 7.10A-C: *Equisetum* (Gametophytes)

A - The male gametophyte is smaller, one cell in thickness and bears the antheridia on the tips of the lobes or on the margin. B – The monoecious prothalli with archegonia and antheridia. C - The male prothalli sustain antheridial growth and the viable gametes are released at maturity.

Archegonium

If the favorable conditions prevail, the first sex organs in *E. arvense* appear when the gametophytes are 30 to 40 days old. Normally archegonia are first to develop in the meristematic margin of the prothallus (Fig. 7.11A). They are situated among the upright green lobes.

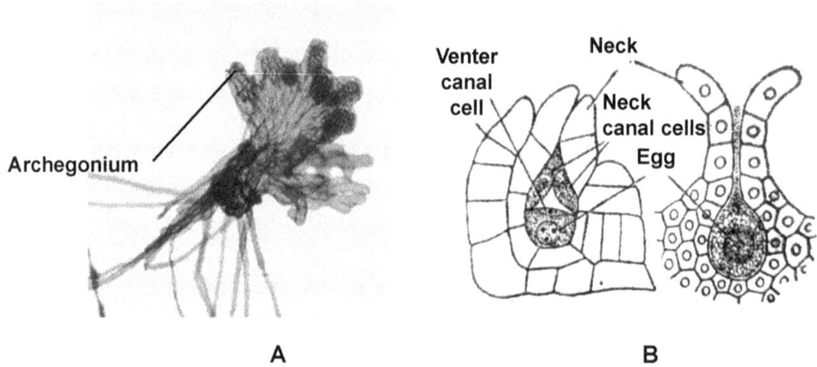

Figure 7.11A,B: *Equisetum* (Archegonia)

A – The archegonia are situated on the meristematic region of the prothallus, at the base of the upright lobes. B- The mature archegonium has neck canal cells, venter canal cell and the egg cell.

Structure of Archegonium

- An archegonium consists of a sunken base in the prothallus and a projecting neck.
- The archegonial neck is composed of four vertical rows of cells where each row is two to four cells in height.
- The terminal cells of the archegonial neck are quite long, which become separated and curved outwards on their maturity.
- The axial row of the cells situated within the archegonium consists of three or four cells - the egg, the venter canal cell and one or two neck canal cells (Fig. 7.11B).
- When there are two neck canal cells, they are boot shaped, e.g., *E. debile*.
- In the last stage of the development all the axial cells of the archegonium disintegrate leaving the canal for antherozoids to swim through and reach the egg, fertilize it for raising the next generation.

Antheridia

In *Equisetum arvense* (Field Horsetail), a monoecious plant, as soon as the formation of archegonia is over, the antheridia develop on the same prothalli. At this stage of antheridial development, the growth of the upright lobes on the prothallus ceases, and the prothallus turns and expands upwards. The antheridia develop in these expanded and turned portions of the prothalli (**PLATE 4P**). However, in starved and crowded prothalli, only antheridia develop. These prothalli may also be called the male prothalli. Thus the antheridia are of two types – **(a)** the embedded type and **(b)** the projecting type. The former type usually develop on the massive tissue of normal gametophytes and the latter develop in the meristematic margin of the starved crowded and dwarfed male prothalli.

- A superficial cell of the meristematic margin of the prothallus acts as an antheridial initial.
- The mature antheridium has a single layered jacket.
- Occasionally there lies a triangular opercular cell at the apex of jacket layer.
- Inside the jacket, the primary androgonial cells divide repeatedly giving rise to a group of cells. Ultimately, the androcyte mother cells are formed.
- Near the nucleus of each mother cell two blepharoplasts appear.
- Each androcyte mother cell divides giving rise to two androcytes.
- The rounded nucleus of the androcyte becomes elongated ultimately becoming crescent shaped. At the same time the blepharoplast becomes spirally elongated.
- Each androcyte metamorophoses into a spirally coiled multi-flagellate antherozoid.

Structure of Antherozoid

- The antherozoids are quite large in size.
- Each antherozoid is spirally coiled with thick and flat posterior end.
- The anterior part is conical and bears numerous flagella.
- The body is almost entirely derived from the nucleus of the androcyte while the numerous flagella at the anterior end of the antherozoid originate from the cytoplasm of the mother cell. The flagella are attached to the blepharoplast, which also lies at the anterior end (Fig. 7.12).

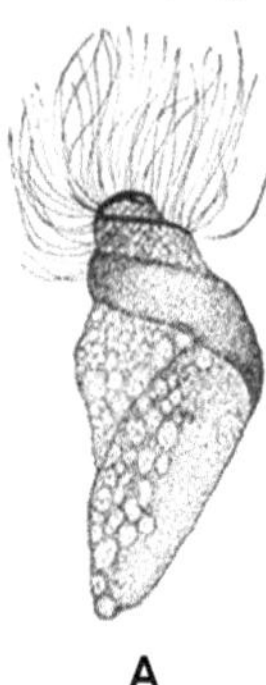

Figure 7.12: *Equisetum* (Antherozoid with apically directed flagella)

Fertilization and Young Sporophyte

The archegonial neck canal cells and the venter canal cell get disorganized making way for the antherozoids to swim down the neck canal and approach the egg. Several antherozoids travel down the neck canal but only one of them unites with the egg forming a diploid zygote. Usually more than one archegonia get fertilized and as a result several sporophytes (8 to 10) develop on the same prothallus (Fig. 7.13). In *E. debile*. Kashyap (1914) recorded as many as 15 sporophytes on a single gametophyte.

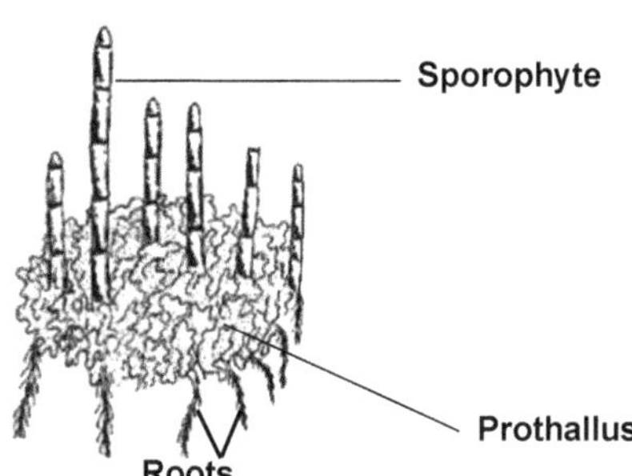

Figure 7.13: *Equisetum* (Large prothallus with six developing young sporophytes)

- As supported by Campbell (1918, 1928); Jeffery (1899) and Walker (1921), usually the first division of the zygote is transverse, giving two cells - the upper, epibasal and the lower, hypobasal cell.
- Both cells take part in the development of the embryo proper and no suspensor is developed unlike in *Selaginella*.
- The next division is at right angles to the first and the quadrant stage of the proembryo is attained (Fig. 7.14A,B).
- Variations in development have also been reported. According to Campbell (1921), in *E. debile* the entire hypobasal half develops into a foot, whereas in *E. arvense* (Sadebeck, 1878) the hypobasal half of the embryo gives rise to both foot and root.
- Later during development, an apical cell with three cutting faces gets organized in the epibasal cell.
- *In *Equisetum debile*, where the entire hypobasal half develops into foot, a superficial cell of the epibasal half lying near the hypobasal half develops into the apical cell of the root which grows downward and reaches the soil.
- The exact point of origin of the stem in *Equisetum* seems to be as indefinite as that of the root. The apical cell that gives rise to it may develop in any one of the four upper octants of the proembryo, usually, however, in the one most closely aligned to the major axis of the archegonium (Fig. 7.14C).
- The stem portion of the embryo grows upright, and emerges through the neck of the archegonium.
- Soon after the stem differentiates into nodes and internodes. The three cells lateral to the apical cell get differentiated into the three leaves of the first leaf sheath. Sometimes there may develop two or four leaves instead of three.
- The stem continues to grow upwards until 10 to 15 nodes and internodes are developed. Thereafter a secondary branch arises from the base of the primary branch. This branch bears four or five leaves at each node (Fig. 7.14D). The growth of the secondary branch is also limited and reaches probably to the same height as that of primary branch.
- More vigorous branches arise successively from the base of the primary stem.
- The secondary branches possess adventitious roots. Several secondary branches of limited growth arise successively from the young sporophyte. All such branches are endogenous in their origin.
- According to Campbell (1928), the origin of secondary branch is endogenous and it arises from the pericycle of the primary root.
- The third or fourth branch grows downward into the soil and becomes the rhizome.
- The portions of the stem and root of the embryo grow and elongate quite rapidly and a young sporophyte is established.

- Thus the life cycle of *Equisetum* is complete (Fig. 7.15).

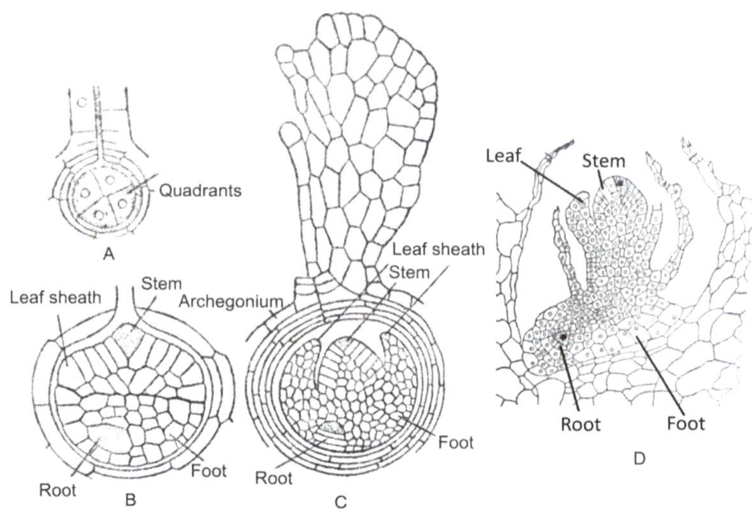

Figure 7.14 A-D: Successive stages in embryo development and sporophyte.

A,B - Development of proembryo. C,D- Young sporophyte with the stem apex, the two first leaf sheaths, the foot and the base of root; but the bud for the secondary shoot is upon the primary root (After Campbell).

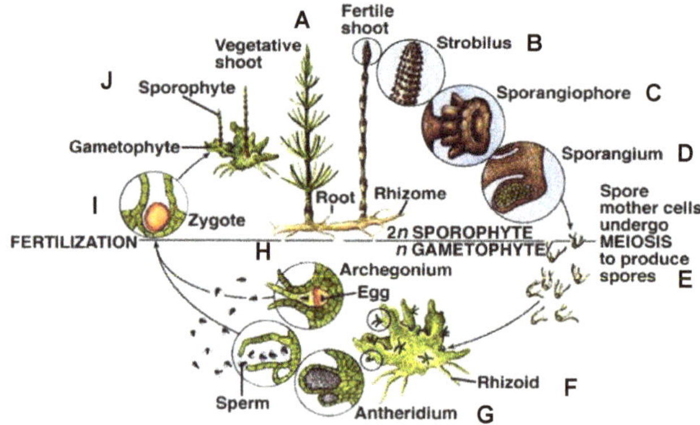

Figure 7.15 A-J: *Equisetum* (Life Cycle)

A, sporophyte; B, strobilus; C, sporangiophore with sporangia; D, L.S. Sporangiphore showing sporangia and peltate disc; E, spore mother cell; F, young prothallus; G, antheridium; H, archegonium; I, zygote; J, young sporophyte.

Class	POLYPODIOPSIDA
Subclass	EQUISETIDAE
Order	EQUISETALES
Family	EQUISETACEAE
	EQUISETUM

QUESTIONS

Q1. Give the salient features of *Equisetum.*

Q2. The *Equisetum* stem is uniquely constructed to exhibit both xerophytic and hydrophytic characters. Comment.

Q3. Elaborate the differentiation of aerial branches in *Equisetum* sporophyte.

Q4. With the help of neat well labelled diagrams, bring out salient features of spore- producing structures in *Equisetum.*

Q5. Write a concise account on nature of prothalli in *Equisetum.*

8 CHAPTER

MARSILEA

Polypodiopsida (after Schuettpelz et al., 2016)

[MONILOPHYTES] (Chase & Reveal, 2009)

[FERNS] (Christenhusz & Chase, 2014)

CLASS POLYPODIOPSIDA (Smith et al., 2006= FILICOPSIDA; **PPG 1, 2016**)

SUBCLASS POLYPODIIDAE (Chase & Reveal, 2009; Christenhusz & Chase, 2014; **PPG 1, 2016**)

ORDER SALVINIALES (Smith et al., 2006; Christenhusz & Chase, 2014; **PPG 1, 2016**)

FAMILY MARSILEACEAE (Smith, 2006; Christenhusz & Chase, 2014; **PPG 1, 2016**)

Water ferns, and heterosporous ferns (incl. "Hydropteridales", Marsileales, Pilulariales).

The members have fertile- sterile leaf blade differentiation; anastomosing veins; aerenchyma often present in roots, shoots, and petioles; annulus absent in sporangia; plants are heterosporous with endosporous germination; monomegaspory seen with gametophytes that are highly reduced.

The Salviniales, includes two families – Marsileaceae and Salviniaceae. As in *Selaginella* and *Isoetes* of the Lycophytes, all members of Salviniales are unique among the leptosporangiate ferns in being heterosporous. Both sporangia types arise (together or separately) within a sporocarp, a rounded, seed-like structure with a hard outer layer. The sporocarp functions in providing protection (being resistant to desiccation) and inducing dormancy under unfavourable conditions. Sporocarps produced at the end of the season will remain dormant until conditions are right for growth which is generally the start of the next growing season. However, in some cases, sporocarps have remained dormant for decades and yet the spores remained viable. Sporocarps open up (= germination) to release spores, which are embedded in a gelatinous exudate (known as a sorophore or massula). As in the

Lycophytes, megaspores are large and produced in few numbers (typically only 1 per megasporangium). The single, haploid nucleus of the megaspore gives rise to a female gametophyte, which may at least partially develop within the megaspore wall (as in the Lycophytes, this process known as endospory). The female gametophyte bears one or more archegonia. Microspores, in contrast, are small and produced in large numbers. Each microspore develops into a male gametophyte, which bears one or more antheridia. Interestingly, heterospory, reduction of megaspore number per megasporangium, and endospory as in *Selaginella* have led to seed habit during evolution of seeds.

SALVINIACEAE [INCL. AZOLLACEAE]–Floating Fern family (type *Salvinia*, after Italian Antonio Maria Salvini, 1633–1729). Includes 2 genera (*Azolla*, mosquito ferns, & *Salvinia*, water spangles)/ca. 21 species. Plants are free-floating, subcosmopolitan, roots present (*Azolla*) or lacking (*Salvinia*); stems protostelic, dichotomously branched; leaves sessile, alternate, small (ca. 1–25 mm long), round to oblong, entire; veins are free (*Azolla*) or anastomosing (*Salvinia*); spores of two kinds (plants heterosporous), large megaspores and small globose, trilete microspores. Spore germination is endosporic.

PEPPER WORT OR FOUR-LEAF CLOVER FERN

Marsileaceae (PPG 1): The Clover Fern family (type *Marsilea*, after Italian Count Luigi Ferdinando Marsigli (1658–1730), Latinized as Marsilius). It includes 3 genera (*Marsilea*, leaves with 4 pinnae, *Pilularia*, leaves filiform or thread-like, and *Regnellidium*, leaves with 2 pinnae)/ca. 61 species.

Rooted aquatics, in ponds, shallow water, or pools, with floating or emergent leaf blades; subcosmopolitan in distribution. Stems are rhizomatous, usually long, creeping, slender, often bearing hairs; leaves occur at widely spaced nodes, leaflets 4, 2 or 0 per leaf (*Pilularia* where the fronds essentially consist of the stems only, any form of flattened laminae having been lost); veins are dichotomously branched but often fuse toward their tips. Sori are borne in stalked bean-shaped sporocarps that arise from the rhizomes or from the base of the petioles, one to many per plant; heterosporous. The sporangia lack annulus. The microspores are globose, trilete, and produce coiled, multiflagellate antherozoids while the megaspores are globose, each with an acrolamella positioned over the exine aperture; perine gelatinous. The spores of Marsileaceae differ markedly from those of Salviniaceae, the only other family of heterosporous ferns and sister group to Marsileaceae, and from spores of all homosporous ferns. The outer spore wall (perine) is modified above the aperture into a structure, the acrolamella. The perine and acrolamella are further modified into a gelatinous layer that envelops the spore. The gelatinous perine layer acts as a flexible, floating structure that envelops the spores only for a short time and appears to be an adaptation to amphibious habitats. The gelatinous nature of the perine layer is likely the result of acidic polysaccharide components in the spore wall with hydrogel (swelling and shrinking) properties.

Megaspores floating at the water/air interface form a concave meniscus, at the centre of which is the gelatinous acrolamella that encloses a "sperm lake." This meniscus creates a vortex like effect serving as a trap for free swimming sperm cells, propelling them into the sperm lake and increasing fertilization chances.

DISTRIBUTION

The genus *Marsilea,* comprising about 58 living species (and 10 fossil species), is worldwide in distribution and very common in warmer part of the world such as tropical Africa and Australia. The species are either aquatic or amphibious (with fruiting bodies in muddy soils) i.e., they grow either completely sub-merged in water, or partly or wholly out of water with their roots embedded in the muddy soil. In India nine species are found, common being *M. minuta, M. aegyptiaca, M. quadrifolia, M. brachypus, M. condenseta, M. gracilenta, M. poonensis, M. brachycarpa, M. coromandelica* while *M. condensata* and *M. rajasthanensis* are near xerophytic. *Marsilea hirsuta,* and Australian species can withstand long dry spells; rest being aquatic or amphibious. *Marsilea minuta* is a very troublesome weed in lowland rice fields and ditches, and can grow under water: during the first half of the growth period of rice plants, it is a severe competitor, especially during the first two weeks after transplantation of young rice plants. It may also affect vegetables that are cultivated in water, such as *Ipomoea aquatica.*

SPOROPHYTE

Morphology

The dominant plant body is a sporophyte differentiated into rhizome, roots and leaves (Fig. 8.1A). The plants produce sporocarps (Fig. 8.1B) which are heterosporous and produce micro- and megasporangia. During sexual reproduction, it produces male and female gametophytes which are haploid. The female gametophyte bears archegonia and the male bears antheridia. Fertilization of the egg cell with antherozoid results in a diploid sporophytic structure – the zygote. Subsequent development leads to a well differentiated diploid plant body.

Rhizome

- It is long and slender and creeps either on the surface like, stolon or slightly below the surface of the soil like, rhizome.
- It grows extensively and is branched with distinct nodes and internodes.
- The branches arise at the bases of petioles and are extra axillary in position, arising in the lateral or oblique position.
- They run in all directions and may get rooted at the nodes. In this way, a single plant may cover an extensive area of about 20 meters diameter or even more.

- The stem remains rhizomatous throughout growth. The internodes are long in the hydrophytic species and short in the sub-terrestrial or xerophytic species.

Root

- The primary root formed in the rhizomatous stem is short-lived and is soon replaced by adventitious roots, which arise gradually at the nodes on the underside of the stem.
- Sometimes roots may also arise at internodes.
- The adventitious roots developing in acropetal fashion are thin and may be branched or unbranched. The number of roots and their size may vary depending on the habitat.
- The roots of *M. minuta* secrete carbonic acid and because of this secretion the roots penetrate the shells of pond snails.

Leaves

- They arise from the upper side of the stem and are arranged in two alternate rows.
- They are long petiolate and compound. The petioles of the submerged plants are long, thin and flexible, with the lamina floating on the surface of water, while the leaves of plants growing in mud or on land have upright, short petioles, with lamina held in a spreading position.
- The lamina is usually divided into four leaflets of the same size. They spring up from the tip of the petiole, so that the leaf apparently looks quadrifoliate (**PLATE 5A, B**).
- Occasionally the number of leaflets may be 5 or even 6 or 8. In outline leaflets are obovate, elliptical or wedge- shaped with entire or dentate margins.
- According to Puri and Garg (1953), the leaf of *Marsilea* is pinnately compound, with four pinnules borne on the slender rachis; two pinnules are noticeably higher than the other two and are inserted on the rachis in alternate fashion.
- A leaflet has many dichotomously branched veins, which are joined with each other by transverse bands and their ends unite to form marginal loops (Fig. 8.1C).
- In *M. minuta*, the pinnae become coarsely toothed. At night the pinnae of aerial leaves become folded upwards and exhibit sleeping movement.
- When young the leaves show circinate vernation (**PLATE 5C**).

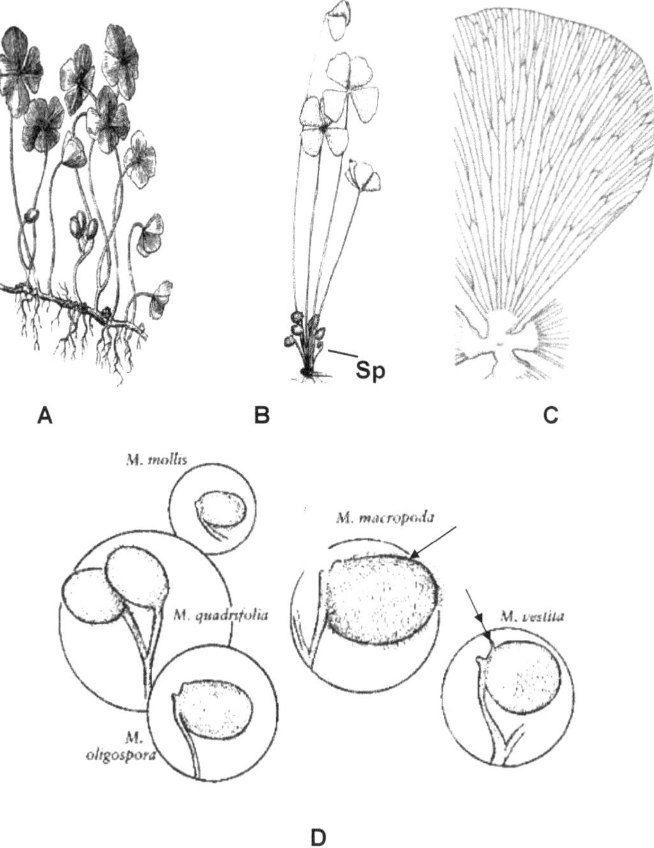

Figure 8.1A-D: *Marsilea* (Morphology)

A – The sporophyte with quadrifolate leaves. B- The sporocarps (sp) arise at the base of the petiole. C – The leaves show dichotomous venation where the veins anastomose at the tips. D- The sporocarps are borne on branched or unbranched petiole. They may be smooth or glabrous (arrow) and bear two teeth-like structures (double arrows).

Sporocarp

- The sporocarps are borne on short peduncles or stalks near the base of the petiole.
- In majority of cases the peduncle of the sporocarp is unbranched and bears a single sporocarp at its apex.
- However, *M. quadrifolia* possesses dichotomously branched peduncle, which bears two to five sporocarps. In *M. polycarpa* a single dichotomously branched peduncle bears about 6 to 26 sporocarps (Fig.8.1D).
- Upon dehiscence, the sporocarp releases an elongate, gelatinous structure, the sorophore, to which are attached several pairs of sori.

Anatomy

Rhizome

- Epidermis is single layered, made up of compactly arranged thick walled cells and lacks stomata.
- Cortex differentiated into three regions – outer cortex is parenchymatous, one to several cells in thickness, sometimes with tannin cells, is also called hypodermis. The middle cortex is formed of aerenchyma where the lacunae are in a ring form and are connected by single layered parenchyma. The inner cortex is solid tissue of several cells thick, filled with starch while outer few layers of inner cortex are sclerenchymatous in nature.
- The vascular cylinder is amphiphloeic siphonostele (**PLATE 5D**).
- It is limited externally and internally by endodermis, called outer endodermis and inner endodermis respectively. Similarly there is outer pericycle and inner pericycle.
- The xylem is in form of a ring, phloem is present on both sides of xylem (Fig. 8.2A, B).
- The deposition of several tissues from cortex inwards is outer endodermis, outer pericycle, outer phloem, xylem, inner phloem, inner pericycle and inner endodermis.
- In submerged species, the rhizomes possess usually a parenchymatous pith, while the plants growing on mud have a more or less sclerenchymatous pith.

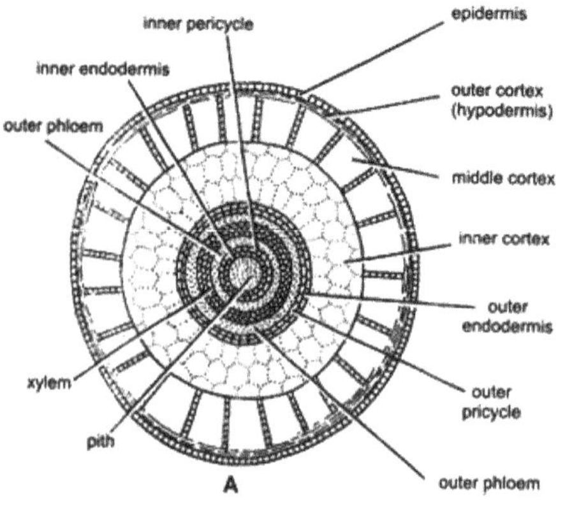

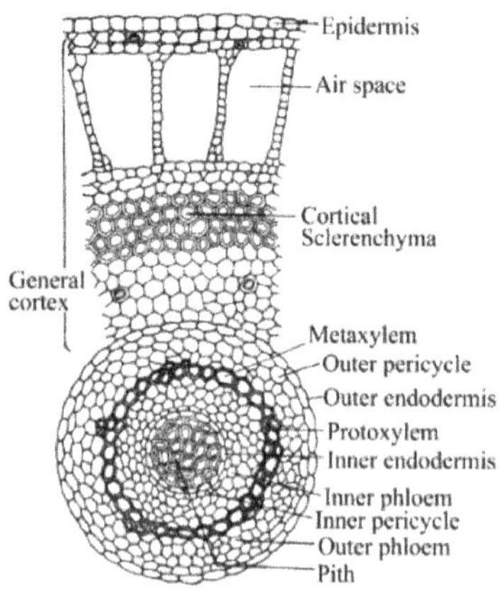

A B

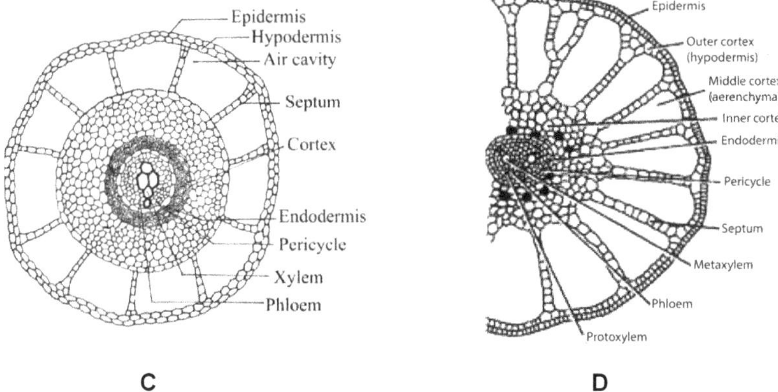

Figure 8.2A-D: *Marsilea* (Anatomy)

A, B –T.S. Rhizome. Outline diagram (A) shows prominent aerenchyma, and amphiphloeic siphonostele in cellular details (B). C –T.S. Root shows aerenchyma and a diarch exarch xylem. D – T.S. Petiole with prominent aerenchyma and a V-shaped stele where the two arms of V are quite separate and somewhat curved from each other.

Root

- Consists of a piliferous layer beneath which lies cortex made of outer and inner regions.
- The outer cortex is aerenchymatous consisting of large air chambers arranged in form of a ring.
- Inner cortex is compact made up of rounded cells containing starch; a few scelerotic cells may also occur.
- Inner cortex is delimited by a single layer of distinct endodermis followed by pericycle.
- Vascular bundle is usually diarch and exarch.
- The xylem is plate-like and occupies the centre of the stele. The protoxylem consisting of two small mass of cells, is towards the periphery (pericycle), while metaxylem is large, plate-like and occupies the centre.
- Pith is absent (Fig. 8.2C).

Petiole

- Single layered epidermis covered by cuticle is made up of rectangular cells.
- The outer cortex consists of few layers of thin walled cells; middle cortex is aerenchymatous consisting of ring of air chambers while the inner cortex is solid, compact tissue several cells deep. The inner layers of inner cortex are filled with starch. Here and there tannin filled cells also occur.

- The stele is covered externally by a single layered distinct endodermis, inner to which lies pericycle.
- The triangular stele lies in the centre and has a single vascular bundle. The xylem is V-shaped, with opening towards the axis i.e., adaxial side.
- The two arms of V are quite separate and somewhat curved from each other. The V-shaped mass of xylem is surrounded by phloem (Fig. 8.2D). The ends of each arm of V-shaped exarch xylem are protoxylem groups whereas the middle regions of the arms of V consist of large tracheids representing the metaxylem groups. The opening of the V is always towards the axis.
- The order of the tissues from inner to outer side of the stele is xylem, phloem, pericycle and endodermis.

Leaflet

- It consists of upper and lower epidermal layers, each made up of a single layer of parenchymatous cells. The continuity of epidermis is interrupted by slightly sunken stomata. Stomata are restricted to the upper epidermis in floating leaves but in species growing on soil, they are present on both the surfaces.
- The ground tissue (mesophyll) lies between the upper and lower epidermis and is differentiated into palisade and spongy parenchyma.
- The palisade part lies just beneath the upper epidermis and consists of columnar cells rich in chloroplasts while the spongy part faces the lower epidermis and consists of rounded cells. In submerged species there is no distinction between palisade and spongy tissues.
- Vascular bundles are embedded in the mesophyll tissue. They are concentric in nature. Each bundle has a central core of xylem surrounded by phloem.

Reproduction

Vegetative reproduction occurs by means of tubers which are small, bud-like with food reserves and arise from rhizomatous stem. These perennating bodies are capable of tiding over unfavourable conditions and germinate under favorable conditions to form new plant.

The other mode is reproduction by spores. *Marsilea,* a heterosporous fern produces special bean-shaped sporocarps outside the water medium. Sporocarp is a bisporangiate structure and produces both micro- and megasporangia.

Sporocarp

- It is a bean -shaped to ovoid, nut or seed-like structure, attached to the basal part of the petiole with the help of a small stalk.
- It is green and soft when young but dark brown at maturity.
- Usually one sporocarp is present at one node but in some species, the number

varies from 2-20. In *M. polycarpa* many sporocarps are attached to one side of the petiole in a single vertical row. In *M. quadrifolia*, pedicels or the stalks are united with one another and are jointly inserted on the petiole. In *M. minuta*, the stalks of all the sporocarps though free, are attached to the petiole at a single point.

- Is differentiated into a stalk and a body. The stalk is fused laterally to the back of the body of the sporocarp, generally forming a distinct ridge called raphe. In some species, the distal (upper) end of the raphe is marked by one or two teeth-like projections or horns, known as tubercles (Fig. 8.1D).
- The wall of the sporocarp is very hard, thick and highly resistant to mechanical injury. It is differentiated into an outer epidermis, followed by hypodermis and an inner parenchymatous zone. The epidermis made up of cuboidal cells, is covered with a thick cuticle. A large number of sunken stomata are present in the epidermis. The hypodermis consists of two layers of radially elongated palisade-like cells, which are compactly arranged. The cells of inner parenchymatous zone, gelatinize forming a gelatinous ring inside the sporocarp.
- Each sporocarp gets vascular supply by a stalk bundle, dorsal bundles, many lateral bundles, placental branches and placental bundles (Fig. 8.3A).
- Sporocarp is supplied by a main dorsal vein which runs along the narrow side facing the peduncle. From the dorsal vein, lateral branches are given alternatively right and left, at right angle to the dorsal vein which supplies laterally. These lateral veins at their middle divide dichotomously. In the region where lateral vein forks, arises a placental bundle which too branches dichotomously. The first and the last lateral veins do not possess placental bundles.
- The sporangia develop in a basipetal manner on the receptacle.

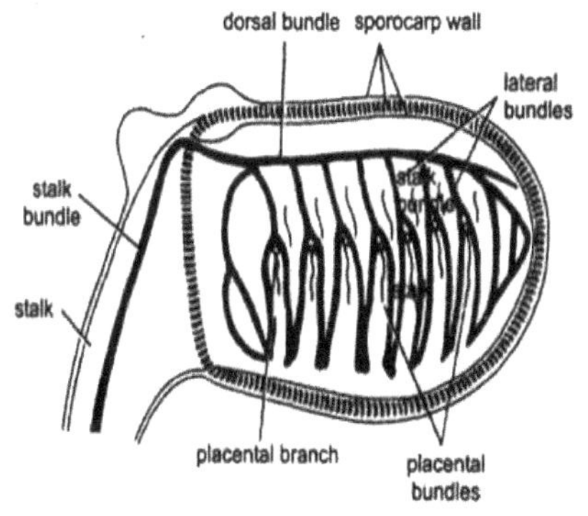

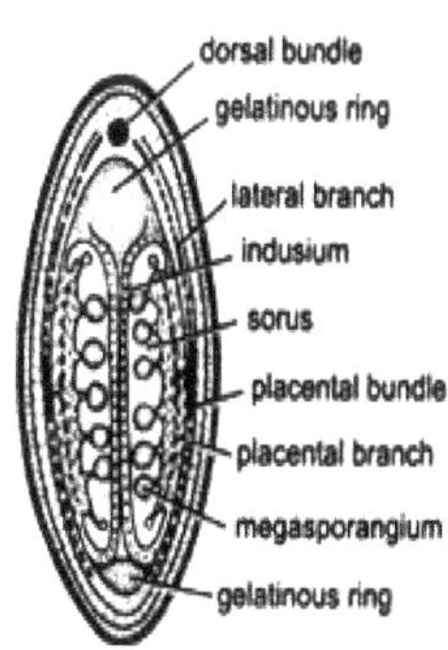

A

B

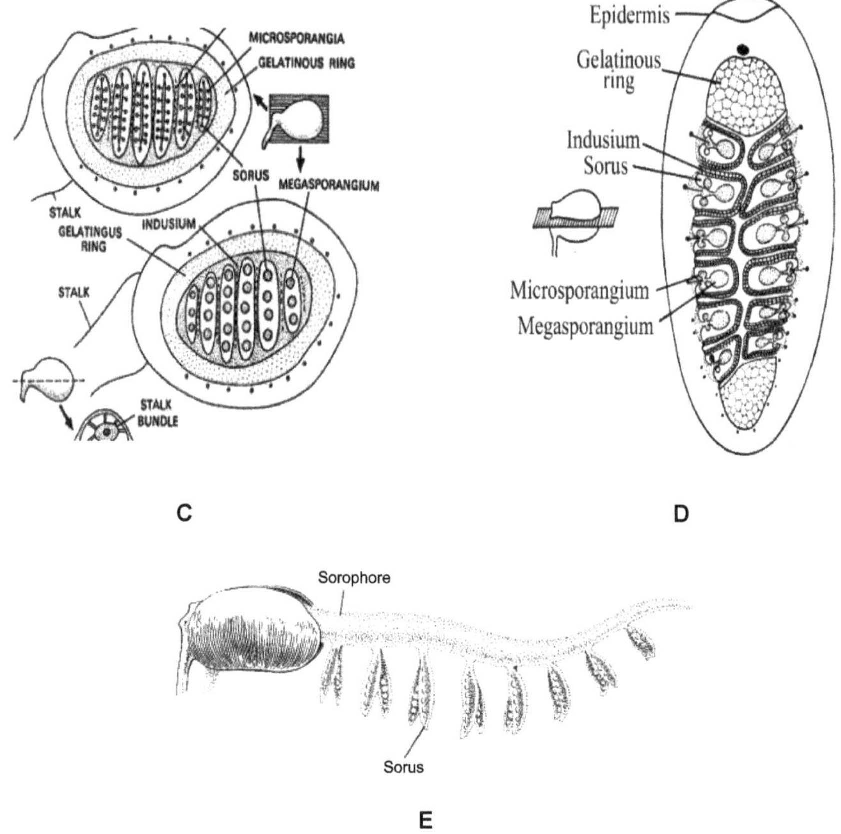

Figure 8.3A-E: *Marsilea* (Sporocarp Structure)

A – A line diagram to show vascular supply to sporocarp. B – VTS sporocarp, C – VLS sporocarp, D – HLS sporocarp. E – W.M. Sorophore to show gelatinous ring, sori and the tip with rudimentary sori.

Sporocarp - structure

The entire internal structure of the sporocarp can be best seen in section cut in three plains: (i) Horizontal Longitudinal Section (HLS.): Section is cut horizontally but the sporocarp is cut longitudinally (ii) Vertical Longitudinal Section (VLS): Section is cut vertically but the sporocarp is cut longitudinally (iii) Vertical Transverse Section (VTS): Section is cut vertically but the sporocarp is cut transversely (Fig. 8.3A-D).

Vertical Transverse Section (VTS)

- In VTS, thick wall with various layers – epidermis, and two layered hypodermis are distinct.

- No peduncle is seen in this view, but gelatinous ring is seen at two poles, the dorsal mass is more prominent.
- Only two sori are seen with indusia on inner side and attached to placental ridge on the outer side.
- Only one type of sporangia seen, all dorsal, placental and lateral bundles seen (Fig. 8.3B).

Vertical Longitudinal Section (VLS)

A VLS of the sporocarp shows the usual wall layers.

- The peduncle along its vascular bundle is cut longitudinally.
- The entire gelatinous ring is cut vertically and it appears as a complete ring around the sori.
- The sori are cut longitudinally, which are arranged in many vertical rows (**PLATE 5E**).
- If the section passes strictly through the median plane of sporocarp, only megasporangia are visible but if it passes slightly away from the median line, only microsporangia are visible (Fig. 8.3C).

Horizontal Longitudinal Section (HLS)

- When we cut a horizontal section (is cut perpendicularly to the flat surface of the sporocarp) of the sporocarp, the wall layers are clear and the peduncle with its vascular system is seen at one end.
- A gelatinous ring appears in the form of a dorsal and a ventral mass at its proximal and distal ends (**PLATE 5F, G**).
- The sori are situated marginally in two alternating rows and remain covered by a two layered indusium (Fig. 8.3D).
- The lateral bundles cut in a transverse section are seen to lie opposite to the base of placentae. Each sorus thus receives a placental supply from lateral bundle.
- The single megasporangium is seen with microsporangia laterally placed.
- The megasporangium has one functional megaspore out of a tetrad of spores developed upon meiosis. The degenerated ones add to the mucilage. In microsporangium all spores are functional and are smaller in size.

Dehiscence of sporocarp

- The spores of *Marsilea* are capable of developing into gametophytes as soon as the sporocarp matures.
- However, under natural conditions being unfavourable the sporocarps do not open until the second or third year of their formation.

- Opening is due to a partial decay of the stony layer of the sporocarp wall. A long delay does not affect the spores as they remain viable for a long time.
- Any time after it is mature, there is an immediate opening of a sporocarp if it is placed in water and a small piece of stony layer is cut away.
- Within 15-20 minutes the gelatinous material within the sporocarp begins to swell because of imbibition of water.
- This swelling causes the two valves of the sporocarp wall to spread apart. As more water is imbibed the gelatinized inner half of the sporocarp wall (the sorophore) begins to project beyond the opened valves.
- Since the sori are attached to the sorophore, they are gradually pulled out as the sorophore elongates.
- A fully elongated sorophore is a long gelatinous cylinder with a length 15 to 20 times that of the sporocarp. On either side of the expanded sorophore is a row of sori, that alternate with one another.
- At the tip of the sorophore are a few small mammilate gelatinous projections, which also alternate with one another. These are rudimentary sori that fail to develop (Fig. 8.3E).
- The indusia are intact and the contained sporangia have an evident jacket when the sori are first freed from the sporocarp wall. There is however, a gelatinization of the indusia and the sporangial jackets five to six hours after extrusion.
- The developing gametophytes generally remain embedded in this watery gelatinous matrix until they are mature.

Morphological Nature of the Sporocarp

Two main views have been put forward by different morphologists to explain the morphological nature of the sporocarp of *Marsilea* - (1) The laminar or leaf segment hypothesis and (2) Petiolar or whole leaf hypothesis.

- The laminar or leaf segment hypothesis. According to the supporters of this hypothesis the sporocarp is a lateral modified fertile segment of the leaf.

Russow (1872) and Busgen (1890) regarded the sporocarp to be made up of 2 leaflets with ventral surface facing each other (ii) Goebel (1882, 1905, 1930) considered the sporocarp as a fertile leaflet (pinna), comparable with one or more leaflets (iii) Campbell (1893, 1928, 1940) regarded the sporocarp as a folded pinnate leaf formed by the fusion of several pairs of pinnae (iv) Bower (1926) stated "the hypothesis seems to be tenable that the sporocarp consists of rachis, bearing 2 rows of pinnules, this is indicated by the venation" (v) Eames (1936) compared it to the tip of the leaf with 4 leaflets. He was of the opinion that the body of the sporocarp represents the two distal leaflets and the region of 1st and 2nd protuberance represent the remaining 2 proximal leaflets (vi) Smith (1938, 1955) considered the sporocarp as a modified pinna with one midrib and several lateral branches (vii) Takhtajan (1953) considered the sporocarp as a more specialised fertile segment of a leaf as in Schizaeaceae. (viii) Puri and Garg (1953) regarded the sporocarp equivalent to a single leaflet which consists of as many pinnules as the number of lateral bundles. (ix) Gupta (1962) regarded the sporocarp to be leaflet with as many lobes as the number of lateral bundles.

- Petiolar or whole leaf hypothesis: Johnson (1898, 1933) proposed this hypothesis and regarded the sporocarp as homologous to the swollen end of the petiole. The marginal cells here develop sporangia instead of lamina. The evidence in support of this theory is that sometimes the primary sporocarp gives rise to secondary sporocarp which develops from the initial arising from the marginal cells of the stalk of primary sporocarp.

GAMETOPHYTE

The microspores and the megaspores both haploid (product of meiosis) germinate to produce micro- and megagametophytes. Early development of the megagametophyte from the megaspore is initially endosporic i.e., within the original spore wall.

Microspore

- A microspore is globose in shape with a pyramidate apex.
- Development of a microspore into a fully mature microgametophyte takes about 12 to 20 hours at ordinary room temperature.
- The microspores germinate immediately after being shed, undergo division forming a small biconvex prothallial cell and a large apical cell.
- The antheridia are directly formed from the apical cell which immediately after its formation divides diagonally to the prothallial cell producing two antheridial initials. Both the antheridial initials then cut off the first jacket cell which does not divide again. Subsequent division results in organization of a second and third jacket cells and primary androgonial cell.
- The primary andrgonial cells remain inside antheridial jacket and each primary androgonial cell divides by four divisions forming 16 androcytes (Fig. 8.4A-F).
- Each androcyte metamorphoses into a cork screw-shaped multiflagellate antherozoid, characterized by the presence of a terminal vesicle.
- A single male gametophyte produces 32 antherozoids with usually one prothallial cell but occasionally two prothallial cells may also be encountered.
- Shortly before the antherozoids are mature, there is a bursting of the spore wall and a protrusion of the antheridia beyond it.
- The antherozoids are liberated by a separation of the jacket cells from one another. Antherozoids of *Marsilea* differ from other ferns in that they have

many more coils, sometimes a dozen or more. They are also unusual in that the flagella are attached only to the broad posterior coils (Fig. 8.4G).

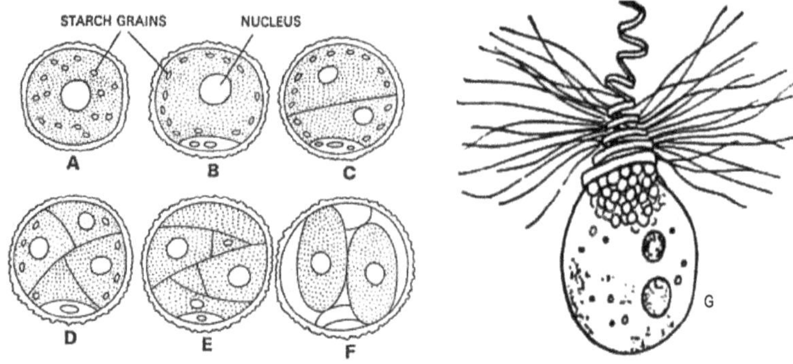

Figure 8.4A-G *Marsilea* (Microgametophyte – Successive stages in development)

Megagametophyte

- Megaspores with two layers - outer, the exospore and inner, the endospore are ellipsoidal with a small hemispherical protuberance at the anterior pole.
- The spore wall about the protuberant portion is relatively thin and has three radiating ridges.
- The protuberant anterior portion of the protoplast consists of densely granular cytoplasm and a small lenticular nucleus; the remainder of the cytoplasm contains abundant starch grains embedded in a watery cytoplasm.
- Development of the megagametophyte to maturity is slightly slower than that of the microgametophyte and takes 14 to 22 hours.
- The first division is transverse and a small nipple-shaped apical cell and a large basal prothallial cell get organized.
- The nucleus lies in the papillate apical cell and it forms the female gametophyte, while prothallial cell with reserves of granular starch, oil globules and albuminous substances, provides nutrition to the developing gametophytes.
- The apical cell further divides by three intersecting vertical walls to establish an axial cell surrounded by three lateral cells. Now the axial cell functions as archegonial initial which divides periclinally to form an outer primary cover cell and an inner central cell. It gives rise to both neck and venter of the single archegonium (Fig. 8.5A-E).
- The prothallial cell does not divide again, only the cells in the region of the apical papillae continue to divide and give rise to the megagametophyte proper.
- The megagametophyte is therefore only a little more than a single archegonium which at maturity breaks through the megaspore wall and protrudes into the sperm lake (**PLATE 5H**).

- Archegonia are typical of ferns with neck and its neck canal cells, and a venter with venter canal cell and an egg cell.
- As the archegonium develops further, the endospore present above the developing megagametophyte is broken and the megagametophyte protrudes through the break in the wall.

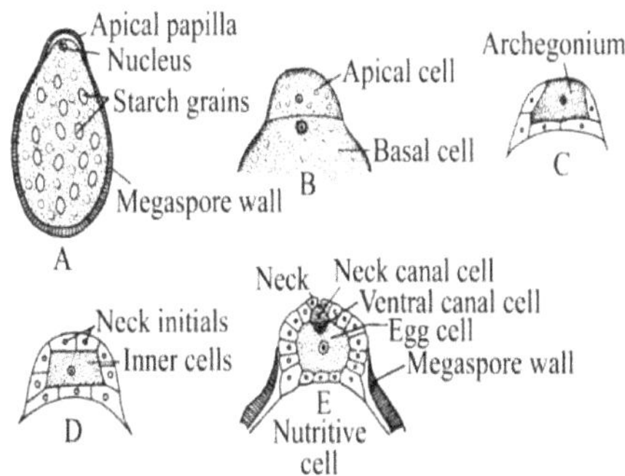

Figure 8.5A-E: Megagametophyte (Successive stages in development)

Fertilization

A mature megagametophyte is surrounded by a broad ovoid gelatinous envelope in which there is more watery funnel shaped portion extending upward from the archegonium. The antherozoids swim into the gelatinous envelope. As the antherozoids swim, the coiled body becomes more lax. Fertilization occurs while antherozoids with posterior end forward enter the mucilaginaous canal formed by disintegration of the neck canal cells and the venter canal cell. The mucilage attracts the antherozoids chemotactically. One of the antherozoids fuses with the egg to accomplish fertilization and form a diploid zygote, thus starting a new sporophytic phase. The zygote develops into a young sporophyte within 2-4 days after fertilization (Fig. 8.6A-C).

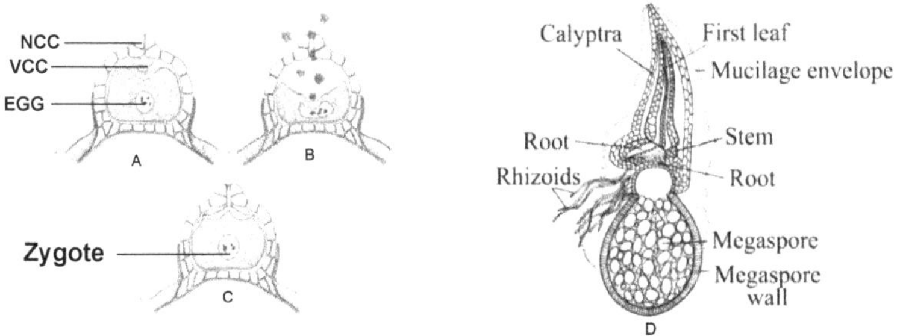

Figure 8.6A-D: *Marsilea* (Fertilization)

A–The egg, venter canal cells and neck canal cells are intact and the neck remains closed. B- The egg becomes cup-shaped and space is created in the neck

through which the antherozoids swim towards the egg. C- Upon fertilization, the zygote becomes spherical and secretes a wall around it. D- Young sporophyte.

YOUNG SPOROPHYTE

The zygote enlarges in size and secretes a thin cellulosic wall around it. It enters into rapid divisions, to give rise to a young sporophyte within 2-4 days. The first formed wall is parallel to the long axis of the archegonium. It divides the zygote into two unequal cells and are referred to as epibasal, the larger and, the smaller hypobasal cells. These two cells undergo transverse divisions to form a quadrant embryo. The cells of the quadrant embryo show definite relation to the primary growth of the embryo. Following divisions result in an octant formation. The root is the first organ to become organized in the embryo. Epibasal half produces shoot and leaf. Hypobasal half produces root and foot. The cell of epibasal half near the neck gives rise to cotyledon and other away from the neck, to the stem. In the same way the cell of the hypobasal half near the neck gives rise to root and other away from the neck, to the foot. Hence, the development is described as lateral. In quadrant embryo the subsequent divisions are irregular. At the time of differentiation of different primordia in the embryo, the adjoining cells of megagametophyte divide periclinally, forming a thick calyptra which encloses and protects the developing embryo. The calyptra is greenish in color. Large number of rhizoids may develop from the base of the calyptra. The embryo gives rise to the young seedling that penetrates through the calyptra and grows into a sporophyte (Fig. 8.6D).

Casual observations on numerous populations of developing megagametophyte of *Marsilea* have shown that some megagametophytes fail to produce sporophytes, but nonetheless they soon turn green and abundant rhizoids arise from the surface of the megagametophyte containing an unfertilized egg. In contrast, the megagametophytes with embryos produce rhizoids almost exclusively at the site on the calyptra adjacent to the development of the embryonic root.

Class	POLYPODIOPSIDA
Subclass	POLYPODIIDAE
Order	SALVINIALES
Family	MARSILEACEAE
	MARSILEA

QUESTIONS

Q1. Explain the morphology of *Marsilea* sporophyte.

Q2. A) Draw T.S. Petiole of *Marsilea*

B) Draw HLS of Sporocarp of *Marsilea*

Q3. Explain the events that occur during dehiscence of sporocarp.

Q4. Trace the embryogeny in *Marsilea.*

PTERIS

Polypodiopsida (after Schuettpelz et al., 2016)

[MONILOPHYTES] (Chase & Reveal, 2009)

[FERNS] (Christenhusz & Chase, 2014)

CLASS POLYPODIOPSIDA (Smith et al., 2006; **PPG 1, 2016**= Filicopsida, Leptosporangiate ferns)

SUBCLASS POLYPODIIDAE (Chase & Reveal, 2009; Christenhusz & Chase, 2014; **PPG 1, 2016**)

ORDER POLYPODIALES (= Filicales in some older literature, Christenhusz & Chase, 2014; Smith et al., 2006; **PPG 1, 2016**)

ORDER POLYPODIALES - THE POLYPOD FERNS

Smith et al. (2006) carried out the first molecular phylogenetic–based higher-level pteridophyte classification. They referred to the ferns (now also include horsetails) as monilophytes, dividing them into four groups or classes (Psilotopsida; Equisetopsida; Marattiopsida; Polypodiopsida), where maximum species belong to class Polypodiopsida. The four-fold grouping has persisted through subsequent systems, despite changes in nomenclature.

Polypodiopsida, a class of Smith et al. (2006) is a rank now used for all ferns (*sensu lato*), and is a subclass Polypodiidae in PPG 1. This group is also informally known as the **leptosporangiate ferns, while the remaining three groups of PPG 1 (=subclasses Equisetidae, Ophioglossidae, and Marattiidae) are referred to as eusporangiate ferns (Fig. 9.1A). The subclass Polypodiidae (after Christenhusz & Chase, 2014; Schuettpelz et al., 2016) has been divided into seven orders,

**Some early classifications used terms Leptosporangiopsida and Filicopsida (class), Filicales (order) and Polypodiaceae (Family) to classify ferns. However, Pichi-Sermolli (1958) dropped the terms Eusporangiatae and Leptosporangiatae 'because according to him, the origin of the sporangia does not offer us a sharply taxonomical distinction'. In his opinion, these terms 'should be used as descriptive morphological terms, not as names of taxa'.

Polypodiales being the largest (Fig. 9.1B). The phylogenetic position of Polypodiales in relation to the other orders of Polypodiidae is shown in the cladogram.

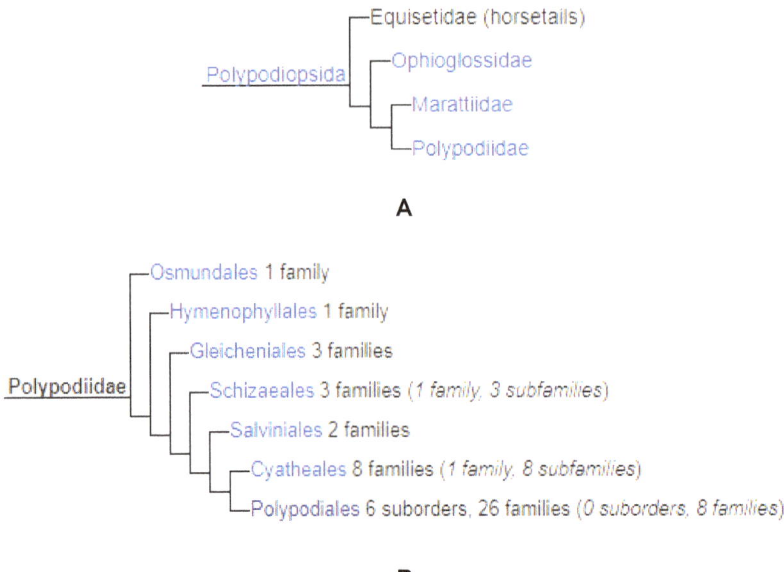

Figure 9.1A,B: Pteridophyte Phylogeny Group

Polypodiopsida has four subclasses (A), in both the Christenhusz and Chase, and the PPG 1 classification, the extant Polypodiidae are divided into seven orders, 44 families, 300 genera, and an estimated 10,323 species (B).

TAXONOMIC CONSIDERATIONS

The order is monophyletic with laterally or centrally attached indusia which may be lost in many lineages); with sporangial stalks 1–3 cells thick, often long; sporangia of mixed stages occur in the sori; sporangia each with a vertical annulus and spores with a coarsely ridged exospore, which is overlain by a thin, sometimes granulate perispore are salient features of the group.

Smith et al. (2006, 2008) divided the Polypodiales into fifteen families. The linear sequence of Christenhusz et al. (2011), incorporated new phylogenetic evidence to include several changes at the familial level making the sequence compatible with the classification of Chase and Reveal (2009), resulting in an expansion to 23 families. Later, classification of Christenhusz and Chase (2014) reduced the number of families recognized in this order to eight. The PPG I classification (Schuettpelz et al., 2016) used a ranking intermediate between the two previous approaches, by introducing a new rank, that of suborder, and organising 26 families (in some cases very narrowly circumscribed) into six suborders, largely returning to the families set out by Christenhusz et al. (2011).

The classifications by Smith et al. (2006) and the Pteridophyte Phylogeny Group (2016), are considered most complete because of phylogenetic acceptance. Table 9.1 summarizes four modern systems of classification; families are listed alphabetically within three broad groups. Although the same families are used in more than one system, circumscriptions may differ. Christenhusz and Chase (2014) used a very broad circumscription of Aspleniaceae and Polypodiaceae, reducing families used in other systems to subfamilies.

Table 9.1: Comparison of alternative subdivisions of Polypodiales

	Smith et al. (2006)	Christenhusz et al. (2011)	Christenhusz & Chase (2014)	PPG I (2016)
Basal families	–	Cystodiaceae	Cystodiaceae	Cystodiaceae
	Dennstaedtiaceae	Dennstaedtiaceae	Dennstaedtiaceae	Dennstaedtiaceae
	Lindsaeaceae	Lindsaeaceae	Lindsaeaceae	Lindsaeaceae
	–	Lonchitidaceae	Lonchitidaceae	Lonchitidaceae
	Pteridaceae	**Pteridaceae**	**Pteridaceae**	**Pteridaceae**
	Saccolomataceae	Saccolomataceae	Saccolomataceae	Saccolomataceae
Aspleniineae eupolypods II (Aspleniaceae)	Aspleniaceae	Aspleniaceae	Aspleniaceae: Asplenioideae	Aspleniaceae
	–	Athyriaceae	Aspleniaceae: Athyrioideae	Athyriaceae
	Blechnaceae	Blechnaceae	Aspleniaceae: Blechnoideae	Blechnaceae
	–	Cystopteridaceae	Aspleniaceae: Cystopteridoideae	Cystopteridaceae
	–	–	–	Desmophlebiaceae
	–	Diplaziopsidaceae	Aspleniaceae: Diplaziopsidoideae	Diplaziopsidaceae
	–	–	–	Hemidictyaceae
	Onocleaceae	Onocleaceae	–	Onocleaceae
	–	Rhachidosoraceae	Aspleniaceae: Rhachidosoroideae	Rhachidosoraceae
	Thelypteridaceae	Thelypteridaceae	Aspleniaceae: Thelypteridoideae	Thelypteridaceae
	Woodsiaceae	Woodsiaceae	Aspleniaceae: Woodsioideae	Woodsiaceae

	Smith et al. (2006)	Christenhusz et al. (2011)	Christenhusz & Chase (2014)	PPG I (2016)
Polypodi-ineae eupolypods I (Polypodia-ceae)	Davalliaceae	Davalliaceae	Polypodiaceae: Davallioideae	Davalliaceae
	–	–	Polypodiaceae: Didymoch-laenoideae	Didymochlaena-ceae
	Dryopteridaceae	Dryopteridaceae	Polypodiaceae: Dryopteridoideae	Dryopteridaceae
	–	Hypodematiaceae	Polypodiaceae: Hypodematioide-ae	Hypodematiaceae
	Lomariopsida-ceae	Lomariopsidaceae	Polypodiaceae: Lomariopsidoide-ae	Lomariopsidaceae
	–	Nephrolepidaceae	–	Nephrolepidaceae
	Oleandraceae	Oleandraceae	Polypodiaceae: Oleandroideae	Oleandraceae
	Polypodiaceae	Polypodiaceae	Polypodiaceae: Polypodioideae	Polypodiaceae
	Tectariaceae	Tectariaceae	Polypodiaceae: Tectarioideae	Tectariaceae

GENERAL FEATURES

The largest order of modern ferns, commonly called the true ferns; also known as Filicales, contains approximately 250 genera and 9500 species. Although members are well represented in the temperate regions, they reach their greatest growth in the moist tropics. They vary in habit from small filmy structures to large tree-like plants. Many species are epiphytic (living perched on other plants) while a number are climbing species and a few aquatic forms are also encountered. Perhaps the most striking species are the tropical tree ferns with their upright, unbranched stems and terminal clusters of large graceful fronds. Members of the Polypodiales differ from members of the other fern orders in being leptosporangiate with sporangium, or spore sac, arising from a single surface cell—and in having small sporangia with a definite number of spores. The wall of the sporangium is almost encircled with a ring of cells, annulus having unevenly thickened walls. The sori may be indusiate (laterally or centrally attached) or exindusiate. The sporangia are distinct with a thin (1-3-celled), generally long stalk, a lateral stomium, and a vertically oriented annulus that is interrupted by the stalk and stomium. When the sporangium is mature, the annulus, acting as a spring, causes the sporangium wall to rupture, thus discharging the spores. The spore germinates giving rise to gametophytes that are green, usually cordate, and surficial. Several of the families contain ornamental cultivars, such as Nephrolepidaceae (*Nephrolepis*, sword fern, boston fern) and Blechnaceae (*Blechnum* and *Woodwardia*). *Matteuccia struthiopteris* (ostrich fern) of the Onocleaceae, has edible croziers.

SELECTED FAMILIES

Aspleniaceae–Spleenwort family (e.g., *Asplenium*, derived from Greek *a*, "without" + *splen*, "spleen," with reference to its use in treating ailments of spleen). It includes 2 genera, *Asplenium* and *Hymen-asplenium* with about 730 species (most in *Asplenium*). The Aspleniaceae consist of terrestrial, epipetric, or epiphytic perennials with rhizomatous stems. The rhizomes are creeping, climbing, ascending, or suberect, and bear clathrate scales at shoot apices and petiole bases. The leaves are monomorphic, simple to multi-pinnate, often with small clavate hairs. There are two back-to-back C-shaped vascular strands at petiole base that fuse distally into an X-shape, venation pinnate or forking, usually free, less often reticulate and without included veinlets. Elongated sori and indusia are present along veins. Sporangia are mixed with stalks 1-row, long. Spores are reniform and monolete, with a winged perine. Some common species are - *Asplenium bulbiferum*, mother fern, with marginal plantlets on leaves; *A. nidus*, bird's nest fern; *A. rhizophyllum*, walking fern, with leaf tips rooting, forming a new plantlet; and *A. trichomanes*, maidenhair spleenwort.

The Aspleniaceae are distinctive within the Polypodiales in possessing shoot apices that bear clathrate scales and leaves with elongate, linear sori and indusia.

Dryopteridaceae–Dryopteroid family (e.g., *Dryopteris*, derived from Greek *drys*, "oak" + *pteris*, "fern," presumably in reference to habitat of taxa in genus). The family includes 26 genera (including *Ctenitis, Dryopteris, Elaphoglossum,* and *Polystichum)* with about 2,115 species.

The Dryopteridaceae like Aspleniaceae consist of terrestrial, epipetric, or epiphytic perennials with rhizomatous stems but with dictyostelic condition. They are creeping, ascending, erect, or scandent to climbing, with non-clathrate scales on shoot apices. The leaves are usually monomorphic, rarely dimorphic, simple or pinnate to decompound; veins pinnate or forking, free to anastomosing, with or without included veinlets. The sori usually round may be indusiate or exindusiate, acrostichoid in some taxa. Indusia, when present, are round-reniform or peltate. The sporangia are mixed, with short to long sporangial stalks in 3 rows. The spores are reniform and monolete.

The Dryopteridaceae are distinct within the Polypodiales with rhizomatous, creeping to climbing plants, the shoot apices with non-clathrate scales, sori exindusiate (acrostichoid in some taxa) or indusiate with the indusia round-reniform or peltate.

Polypodiaceae–Polygram/Polypody family (e.g., *Polypodium*, word derived from Greek *polys*, "many" + *pous*, "foot," in reference to knob-like petiole bases left after leaf abscission). It includes approximately 65 genera and 1,652 species.

The Polypodiaceae consist of epiphytic (usually), epipetric, or terrestrial perennials. The stems like most members of the order are rhizomatous. The rhizomes which are dictyostelic, long to short-creeping, bear scales. The leaves are simple (unlobed to pinnatifid) to 1-pinnate (rarely more), monomorphic or dimorphic; blades may be glabrous or possess hairs or scales. The leaves in many taxa abscise near the base, leaving short petiole bases (termed "phyllopodia"). The veins are often anastomosing or reticulate, sometimes with included veins, or free. The sori are abaxial (rarely marginal), round, oblong, or elliptic, rarely elongate or acrostichoid, the receptacle often with paraphyses, exindusiate (in some taxa covered by caducous scales when young). The sporangia are mixed, with stalks in 1–3 rows, often long while the spores are hyaline to yellowish, reniform and monolete (non-grammitids) or greenish and globose-tetrahedral and trilete (most grammatids). The Polypodiaceae, as treated by Smith et al. (2006), also includes about 20 genera of the so-called grammitid ferns (incl. *Grammatis*) and 600 species of mostly small, tropical epiphytes with simple leaves. The grammatid ferns are often treated as the family Grammitidaceae nested within the Polypodiaceae.

The Polypodiaceae are distinctively characterized within the Polypodiales as exindusiate, mostly epiphytic ferns with sori that are usually round, oblong, or elliptic, rarely elongate or acrostichoid.

Brake Ferns

Pteridaceae– Commonly known as Pteroid fern family it is nested within suborder Pteridineae with 53 genera and about 1,211 species. They consist of terrestrial, epipetric, or epiphytic plants, rarely floating aquatics (*Ceratopteris* spp.). The stems are rhizomatous, and the rhizomes are creeping to erect, bearing scales or hairs. The leaves are simple, pinnate, pedate, or decompound, veins free or anastomosing. The sori are exindusiate, either marginal with a false indusium formed by a reflexed marginal flap or intramarginal, in lines along the veins, and the receptacle which is generally not raised. The sporangia are mixed, with stalks 1–3 cells thick, often long. The spores are globose or tetrahedral, trilete, and ornamented.

The Pteridaceae comprises five monophyletic groups: Cheilanthoideae (including *Cheilanthes, Dryopteris, Gaga, Pellaea*, and *Pentagramma*); Cryptogrammoideae (*Coniogramme, Cryptogramma*, and *Llavea*); Parkerioideae (*Acrostichum* and *Ceratopteris*); Pteridoideae (*Pteris, Taenitis*, and 11 other genera); and Vittarioideae (*Adiantum* and 11 other genera, incl. *Vittaria*).

PTERIS

Pteris is a large fern genus belonging to the family Pteridaceae subfamily Pteridoideae, distributed in tropical and subtropical regions. *Pteris,* estimated to include about 250 species in the world is a species complex making it taxonomically an interesting taxon. The monophyly of the genus *Pteris* has been confirmed using six cpDNA and one nuclear marker. Furthermore, *Pteris* is classified into three subgenera and 16 sections. Many members are known by the overall term 'brake fern' but some are described more specifically – *Pteris cretica*, for example, is known commonly as a Cretan brake.

Most species of *Pteris* grow in warmer areas, but some are also found in cold temperatures. They grow either terrestrially or lithophytically on rocks in shaded canopy, and open areas, forests, coastal, and xeric niches. Some species also survive in soil contaminated arsenic or other metals and are known to be hyperaccumulators.

The *Pteris* species are easily distinguished from their sori characters, sori being linear and located in margins of leaves, but usually do not reach the apices of segments. The sori are protected by a false indusium. The spores on germination give rise to a prothallus, the gametophyte. The gametophyte development of several *Pteris* species including, *P. vittata, P. finotii, P. fauriei, P. exelsa, P. wallichiana, P. ensiformis, P. cretica, P. multifida, P. deflexa, P. denticulata, P. tristicula, P. faurirei, P. incompleta, P. berteroana, P. chilensis*, and *P. tripartita* are reported having some unique characters.

In *Pteris vittata* sori are arranged in unbroken, copper-coloured ridges along both margins of the frond. The dust-like spores upon release germinate to form minute, fragile, green, membranous gametophytes called prothalli. The bisexual gametophyte usually grows close to the soil, producing female (archegonia) and male (antheridia) sex organs on its lower surface. Once the sex organs are mature, and water is available, the antherozoids move towards archegonia, fertilizing the eggs resulting in zygote formation. Further development leads to embryo which grows into a new sporophyte. The lifecycle is complete, with an alteration between two morphologically distinctive stages.

Given the ease with which *Pteris vittata* grows across a wide-range of anthropogenic habitats, and its natural ability to accumulate arsenic, besides complex biological interactions with microbes, it has been intensively investigated as an 'environmental cleaner'. Trials for significant commercial applications are on contaminated agricultural and industrial soils, and on waste resulting from mining activity. In addition to heavy metal accumulation, *Pteris* foliage can be harvested several times a year, without diluting either the fern's health or its capacity to accumulate arsenic. Roots or rhizoids of *P. vittata* are likely to be the main location of arsenate reduction, with arsenite form preferentially loaded into the xylem.

DISTRIBUTION

Pteris is a genus of robust ferns which grows naturally in tropical or sub-tropical climates in various areas of the world. This fern is reportedly a 'species complex' and includes five cytotypes, viz. diploid, triploid, tetraploid, pentaploid and hexaploid with the basic number being 29 chromosomes. The common Indian species are: *Pteris vittata, P. cretica, P. biaurita, P. quadriaurita, P. stenophylla and P. wallichiana.*

Pteris tripartita is an herbaceous fern native to tropical areas in Africa, Asia and Oceania. It is argued occasionally that *P. tripartita* might be native to the New World because it occurs in remote areas of Suriname. In the New World it is naturalized at scattered localities from Florida to northern South America, including the Caribbean. It is reported from Africa, Asia, North America, Central America, the Caribbean, South America and Oceania. The species can grow up to 2 metres in height.

Another species, a quick-growing fern *Pteris vittata* though originated from Asia, is widely adapted throughout the tropics and subtropics. In its native habitat it favours open sites on limestone, but in urban environments, it is found growing well on walls and in concrete. *Pteris vittata* adapted to a variety of soils, has ability to accumulate arsenic and has a potential for phytoremediation or phytoextraction.

The Pteris members thrive in light, loamy soil with plenty of space for root growth. In parts of the United States, certain species have grown so prolifically that they have been classified as invasive weeds. Types of *Pteris* vary from those requiring greenhouse conditions with a minimum temperature of 55ºC to those which can be grown outdoors. Indoor *Pteris* plants such as *cretica* species can easily be grown as houseplants. They need to be kept hydrated and placed out of drought and away from sources of dry heat such as radiators. Ladder Brake is a popular ornamental species, and is readily available from nurseries. It prefers an alkaline soil and can tolerate more sun and drier conditions than many ferns.

Morphology

- *P. vittata* Ladder Brake; Chinese Brake grows in clumps that rarely extend to more than 12 inches in height or width **(PLATE 6A,B)**.
- It is a perennial herb with short-creeping, densely scaly stout rhizome.
- The leaves of fern are often called fronds.
- It grows in wild and is also cultivated as a garden plant because of its attractive form.

Roots

- Fern roots are generally thin and wiry in texture and grow along the stem (Fig. 9.2A). The primary root is ephemeral, and is replaced by a large number of adventitious roots developed all over the surface of the rhizome. The roots are small and branched. They absorb water and nutrients and help secure the fern to its substrate.
- The adventitious roots are produced secondarily from the tissues of the stem especially the base of each petiole.
- The roots are slender and profusely branched, so that an old rhizome becomes densely covered by a mass of adventitious roots (Fig. 9.2B, **PLATE 6C**).

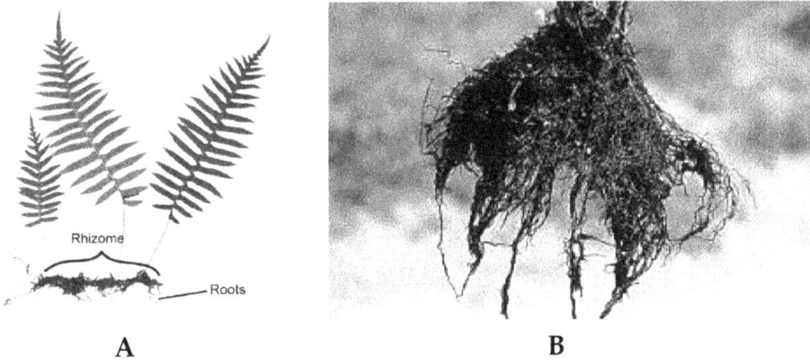

Figure 9.2A,B: *Pteris* (Morphology)

A- The whole plant arises from the rhizome which remains covered with adventitious roots and scales. B – The adventitious roots are well developed at maturity and cover the rhizome.

Fronds/ Leaves

- Leaf shape, size, texture and degree of complexity vary considerably from species to species.
- Fronds are finger-like and remain clustered near the apex of the rhizome (Fig. 9.3A).
- Fronds are usually composed of a leafy blade and petiole (leaf stalk) (Fig. 9.3B).
- Petiole is green to pale brown, 1-30 cm, densely scaly; scales extend to and along rachis which is not winged.
- Scales of rachis grade into uniseriate hairs or ramenta on abaxial costae, hairs are absent on abaxial costae; proximal pinnae not divided or lobed.
- The midrib is the main axis of the blade, and the tip of the frond is its apex.
- The blade may be variously divided, into segments called pinnae; single leaflets are pinna (Fig. 9.3B). Pinna may be further divided, the smallest segments are pinnules (Fig. 9.3C).
- The leaves are bipinnate in *P. biaurita*. Leaf apex is occupied by an old leaflet or pinna (Fig. 9.3A, B).
- Pinnae are smaller near the base larger near the middle, and again smaller towards the apex of the leaf. Every pinna is traversed by a central midrib which gives off lateral veins which are free, open, forked (Fig. 9.3D).
- During reproductive period, narrow sori are located along the margins of the pinna of the fertile leaf or sporophyll (Fig. 9.3E, **PLATE 6D**). The margin of the pinna folds, partly covering the sori with a false indusium.
- The young leaves are circinate and take about two years to reach maturity. In the first year, the petiole, rachis and the leaflets remain coiled. Such young leaves are called circinate leaves (Fig. 9.3F, **PLATE 6E**). As new

fronds emerge, generally in the spring, they unroll. The unrolling fronds are called fiddleheads.

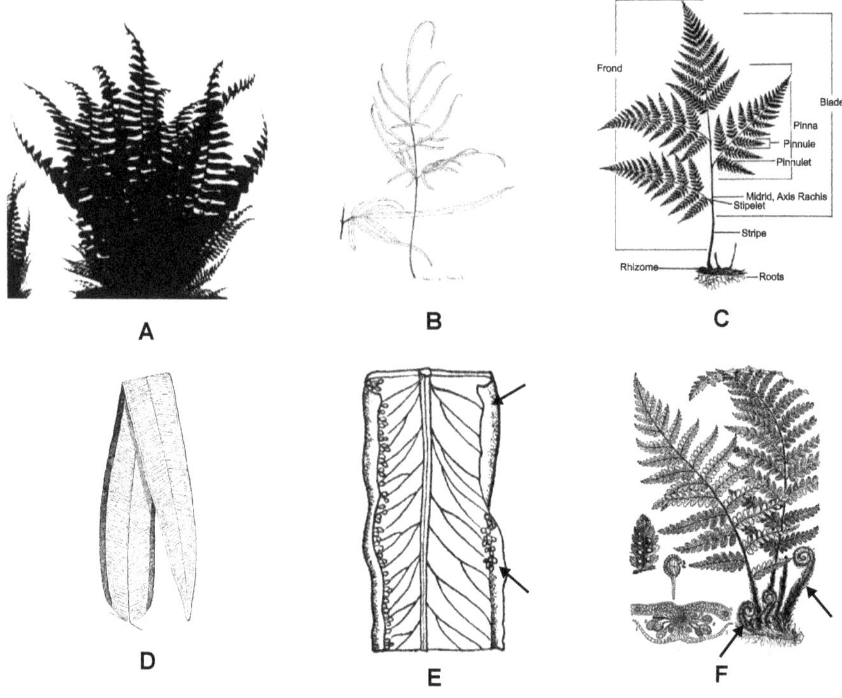

Figure 9.3A-F: *Pteris* (Morphology)

A, B- Leaf shows a long petiole and rachis. In B- Leaf is unipinnate with an odd pinna at the tip. In C- Bipinnate leaf is seen with organization into pinna and pinnules. D- Shows a well-defined midrib and open furcate venation in the blade. E- On the margin sori (arrow) are also distinct. In F croziers (arrows) can be seen. These are the young leaves which keep the apical region coiled so as to protect it from any damage and slowly unroll as they grow.

Anatomy

Root

- An outer piliferous layer or the epidermis is single layered and some cells give rise to root hairs. It is followed by a cortex and a central stele.
- The cortex is differentiated into an outer parenchymatous cortex and an inner sclerenchymatous cortex.
- The endodermis and pericycle are inner to cortex, and enclose a protostele with diarch and exarch xylem. Phloem surrounds xylem (Fig. 9.4A, **PLATE 6F**).

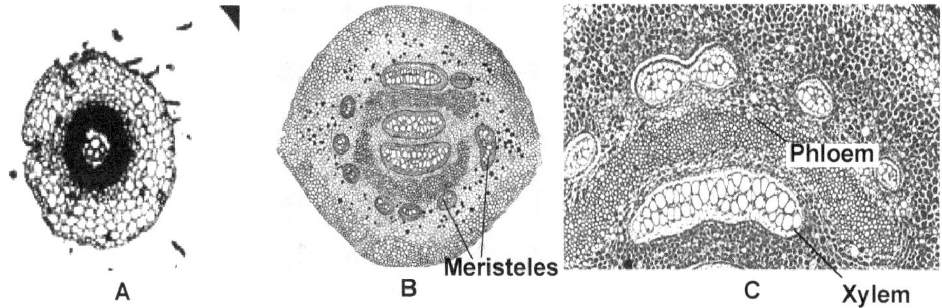

Figure 9.4A-C: *Pteris* (Anatomy)

A- T.S. Root showing epidermis with root hairs, broad outer cortex and sclerenchymatous inner cortex which encloses a stele. B – T.S. Rhizome showing polycyclic dictyostele with meristeles embedded in the ground tissue. C- A meristele enlarged to show a band or plate-like mesarch xylem surrounded by phloem.

Rhizome

- The outermost layer is epidermis followed by a well differentiated cortex.
- The outer cortex may be sclerenchymatous while the inner cortex which lies interspersed among meristeles is parenchymatous and is known as ground tissue (Fig. 9.4B).
- The stelar anatomy of the rhizome which varies from species to species, and sometimes even in same species, is rather complicated and is very variable. Usually, in the case of creeping rhizomes, the vascular cylinder is a solenostele, but in the species with upright rhizomes, it is a polycyclic dictyostele with prominent overlapping leaf gaps.
- In general, the stele is made up of a number of meristeles arranged in two rings. While the inner ring consists of 2 to 3 large meristeles, the outer ring comprises a number of small meristeles (**PLATE 6G**).
- Each meristele has a band or plate-like mesarch xylem surrounded by phloem made up of only sieve cells and phloem parenchyma. It completely surrounds the xylem (**PLATE 6H**).
- Xylem present at the centre of meristele shows central protoxylem surrounded on either side by metaxylem (Fig. 9.4C). It consists of tracheids and xylem parenchyma.
- Each stele is bounded by its own endodermis, inner to which lies a pericycle (Fig. 9.4B, C).

Pinna

- A typical fern pinna is dorsiventral with the mesophyll cells differentiated into palisade parenchyma located under the upper (adaxial) epidermis and spongy parenchyma above the lower (abaxial) epidermis.

- *Pteris vittata*, however exhibits undifferentiated mesophyll consisting of loosely arranged lobed parenchyma cells with abundant intercellular spaces.
- The stomata are restricted to the lower epidermis which has smaller cells and the more sinuous walls.
- The main vascular bundle of the pinna is triangular (**PLATE 6I**).

Rachis

- The outermost is epidermis with a thick cuticle and is followed by hypodermis and ground tissue, the cortex (Fig. 9.5A).
- The vascular bundle is V-shaped with the open side and the notch of the V aligned with the central groove on the adaxial surface (Fig. 9.5A, **PLATE 6J**).
- The xylem strands in each of the lateral arms of the V-shaped vascular bundle have a distinct sea horse outline with their distal adaxial ends curved into pronounced hooks that turn inward facing toward the center of the rachis (**PLATE 6K**).
- Vascular bundles with such xylem configuration have been termed hippocampus vascular bundles by Ogura (1972).
- Vessels have been reported in rachis of pteroid ferns viz. *Ampelopteris prolifera* Copel and *Thelypteris interrupta*.
- The hairs are also seen to arise from outermost layer-epidermis.

Petiole

- In a cross section petiole shows a monostratified epidermis covered by a thick cuticle and 2-18 layers of sclerenchyma and parenchyma that represent a cortex.
- In large plants as in *P. deflexa* the whole cortex is sclerified. This subepidermal supporting tissue has gaps along the petiole formed by ventilation area. The presence of tannins is common in cortex which gives mature petiole dark brown colour.
- The distinguishable feature of the petiole used as a taxonomic indicator is number and position of grooves, which in most ferns occur on the adaxial surface and may be found on the lateral sides as well.
- Generally petioles have from one to three grooves which may start in any portion of the petiole and run on into the rachis.
- In *Pteris vittata* while petiole is adaxially sulcate with a single central groove that is confluent with the groove of the rachis, the abaxial portion has a uniform convex outline.
- Besides grooves, the number, shape, and the organization of vascular bundles, and the configuration of the xylem strands within each vascular bundle have been used as a taxonomic character.

- A considerable evolutionary importance is also attached to the form and arrangement of the vascular tissue in petiole (Foster & Gifford, 1989).
- Each vascular bundle is surrounded externally by the monostratified endodermis and a pericycle consisting of 2-3 cell layers surrounding phloem and xylem.
- In *P. cretica* and *P. multifida* there are two bundles on the base that fuse in the lower third of the petiole by the V shaped abaxial ends. Most species however, have monosteilic petioles along the entire shaft. The stele is V-shaped in *P. ensiformis, P. cretica,* U- shaped in *P. vittata, P. mulifida* while an inverted Ω shaped in *P. deflexa, P. inermis.*
- In *P. vittata,* the U-shaped vascular bundle, open side of which is oriented toward the adaxial side of the petiole is seen (**PLATE 6L**). However, the abaxial portion of the strand is consistently convex and follows the contours of the petiole. Like rachis, a hippocampus xylem strand also occurs with slight variation that involves divergence of the basal ends of the strands facing the abaxial surface (**PLATE 6M**).
- Xylem is characterized by having curved ends in *P. vittata* or long bent ends as in *P. denticulate.* Around the vascular bundles strands of fibres consist of 4-12cells which are isolated in small plants.

  ***Depending on the shape of vascular strands and xylem structure, a classification of vascular bundles has been proposed:*

  TYPE 1- V shaped with the ends of xylem folded in a short hook with three protoxylem areas eg P. cretica

  Type II- U shaped with either shortly curved xylem ends with 5-6 protoxylem areas (Type IIa) as in P. vittata and Type IIb with the ends long extended sometimes joined to the main axis with 4 protoxylem areas as in P. denticulate (Fig. 9.5B-E).

  Type III inverted Ω with more than ten protoxylem areas and the xylem interrupted by parenchyma bands P deflexa.

Whatever is the type, the xylem in all cases is always mesarch with protoxylem of helical to ringed surface and metaxylem with long lenticular pits.

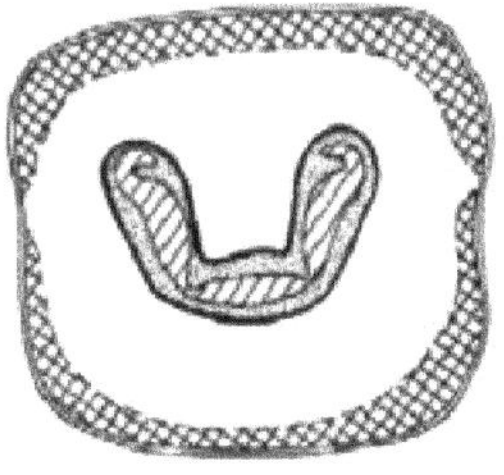

A

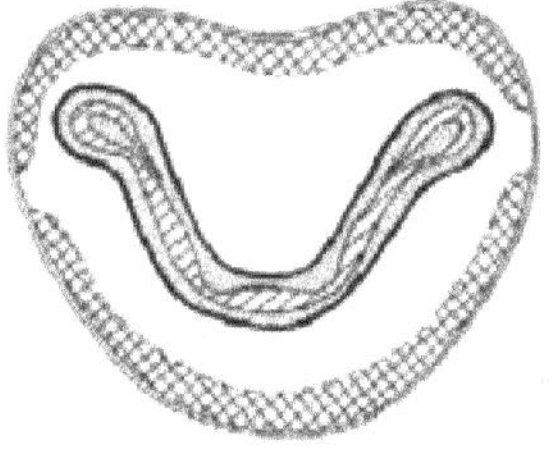

B

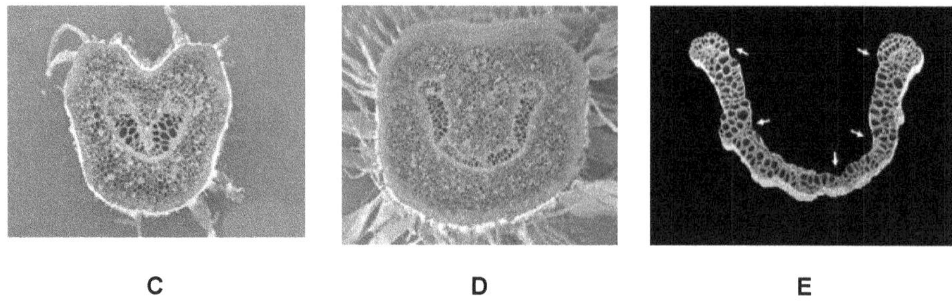

Figure 9.5A-E: *Pteris* (Anatomy)

A, B - line diagrams. C- T.S. rachis to show V-shaped stele. D- T.S. Petiole with U shaped stele, shortly curved xylem ends with 5-6 protoxylem areas (Type IIa) as in *P. vittata.* The ends of xylem appear sharply curved as in rachis (E).

Sporophyll

- Leaf bearing sori is called as sporophyll.
- The sori present on the lower surface are linear, submarginal, and are continuous, commonly known to form a coenosorus (**PLATE 6N**).
- Sori are exindusiate, either marginal with a false indusium formed by a reflexed marginal flap or intramarginal in lines along veins (Fig. 9.6A, B), the receptacle is generally not raised (**PLATE 6O**).
- Nearly all pinnae in mature plants are fertile except the reduced basal ones. The portion of the leaf surface to which sporangia are attached is known as a receptacle.

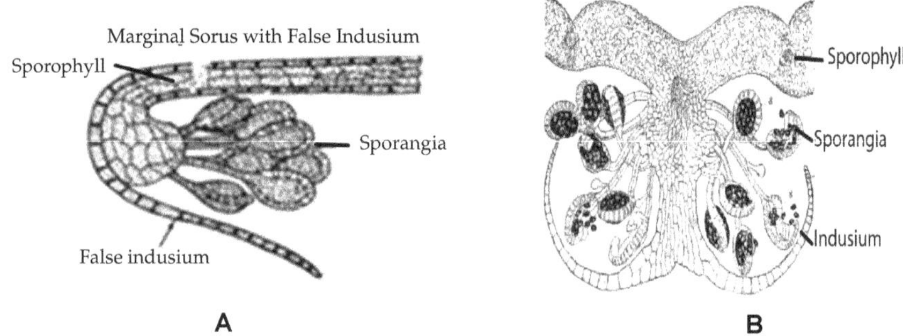

Figure 9.6A,B: *Pteris* (sporophyll)

V.S. Sporophyll of *Pteris* to show false indusium in (A) and true indusium of *Dictyopteris* in (B).

Reproduction

Sporangia

- Sporangia of a sorus are in mixed stages of development, with stalks 1–3 cells thick (multicellular), often long and a capsule which is of the shape of a biconvex lens.
- The wall of the capsule is composed of a single layer of thin walled cells.
- A row of specially thickened cells, known as annulus partially surrounds the capsule. At one side of the annulus, there is a stomium composed of thin walled cells (Fig. 9.7A).
- On maturity, the sporangium contains about 32 to 64 dark, ragged walled spores in it.

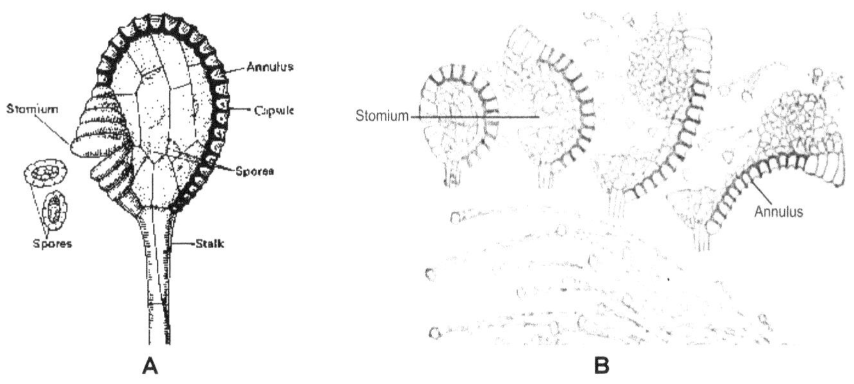

Figure 9.7A, B *Pteris* (Reproduction)
A-Sporangium in V.S. shows stomium with thin walled cells and annulus with thick walled cells, which help in release of spores contained in the capsule, the sac-like structure. B- Spores are released by coordinated action of stomium and annulus.

GAMETOPHYTE

Spores

- The haploid structures are produced through meiosis in spore mother cell.
- They are globose or tetrahedral, and ornamented, and are bluntly or roughly triangular with a distinct triradiate mark.
- The spore wall is thick and has an outer exine and an inner intine; perispore (seen in *Equisetum)* being absent.
- The spores vary in size with the species.
- The members of the family have evolved an ingenious cavitation catapult to disperse their spores.

- The mechanism relies almost entirely on the annulus, a row of 12–25 cells, which successively: (i) stores energy by evaporation of the cells' water content, (ii) triggers the catapult by internal cavitation, and (iii) controls the time scales of energy release to ensure efficient spore ejection.
- Upon reaching maturity, the sporangium is exposed to dry air allowing water to evaporate through the thin outer walls of the annulus cells.
- The architecture of the cells is such that the decrease in cell volume forces the thick radial walls to rotate towards each other.
- This rotation of radial walls drives the tearing of the sporangium and the annulus bends backwards. The force required to bend the annulus wall during opening is balanced by the negative pressure or water tension that develops inside the cells (Fig. 9.7B).
- When the water tension is too large, cavitation occurs, and bubbles are formed within several cells. Without a continuous column of water to sustain the elastic forces in the annulus walls, the elastic energy is quickly released, leading to fast closure and ejection of the spores as in a catapult (**PLATE 6P**).

Germination of spore

- The exine ruptures and the spore grows into a small cylindrical thallus cell.
- A unicellular rhizoid arises from the thallus cell and develops into long and tubular, primary rhizoid.
- The thallus cell divides transversely to form a short filament of green cells. Soon an apical cell gets organized.
- The cells behind the apical notch divide along two or more planes and form a thick central cushion. The secondary rhizoids arise from the posterior regions of the prothallus.
- The single apical cell gets replaced by two or more marginal initials. Soon a heart shaped prothallus is formed with a cushion which is multicellular and wings which are thin and one celled thick (Fig. 9.8A).
- The prothallus may be monoecious or dioecious. In the former case, the sex organs are formed on the lower surface. The antheridia arise in the basal region of the prothallus (**PLATE 6Q**) and appear first. The archegonia are formed later and just behind the apical notch (Fig. 9.8B, **PLATE 6R**).
- The antheridia which are not embedded structures have a single layered jacket that encloses a mass of antherozoids.
- The archegonia have a venter embedded in the prothallus while the neck protrudes out. Just before fertilization, the venter canal cell and the neck canal cells disintegrate giving rise to a mucilaginous mass that facilitates

antherozoid movement towards the egg and fertilization ensues resulting in embryo formation (**PLATE 6S**). Later a new sporophyte is produced (Fig. 9.8C, **PLATE 6T**) and as it becomes independent, the prothallus which bears it, gets disintegrated.

- However, a number of *Pteris* species exhibit another way to produce sporophytes: they arise out of gametophyte cells without fertilization, in a process known as apogamy. Many characteristics have been used to infer the occurrence of apogamy, e.g., ploidy levels, spore number per sporangium, spore size, the development of gametangia, and the formation and morphology of young sporophytes.

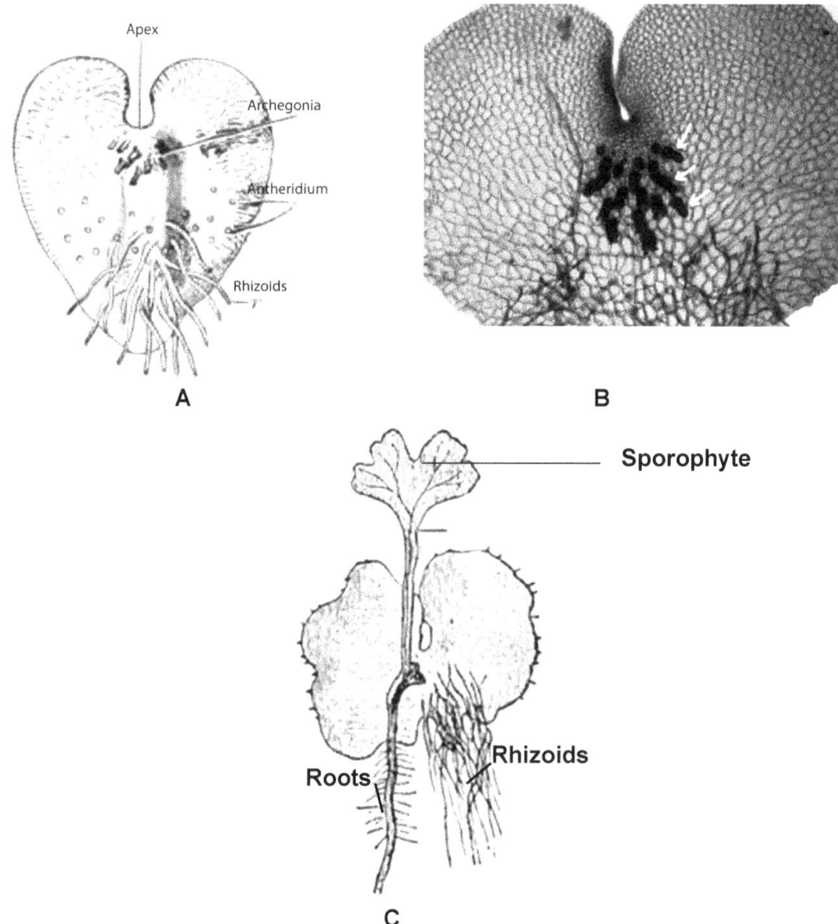

Figure 9.8A-C: *Pteris* (Prothallus)

The heart shaped prothallus bears rhizoids and sex organs on the under surface (A). In (B) prothallus with archegonia (arrows) near the apical end is seen. A young sporophyte is seen in (C) which soon forms its root and becomes independent.

Young sporophyte

- In *Pteris* the first division of the zygote is vertical but in plane at right angles to the long axis of the prothallus. The embryo thus has an epibasal half lying towards the apex of prothallus and a hypobasal cell lying towards the base.
- This division is followed by a second transverse division resulting in the formation of a quadrant. Later a 32-celled embryo is formed due to further divisions of the quadrant.
- The differentiation of embryo begins at this 32-celled stage. No suspensor is formed; the hypobasal cells form first root and foot, while epibasal cells form cotyledon and shoot apex.
- The foot is embedded in the gametophytic tissue and absorbs the nutrients for the developing sporophyte. Soon the prothallus exhausts its nutrients and the young sporophyte becomes independent.

**P. vittata* has a chlorophyllous gametophyte of limited growth, the most vulnerable phase of the life cycle (Zhang et al., 2008). The appearance of AM fungi *Glomus intraradices* at the gametophyte phase may significantly shorten the period when the small plants are especially susceptible to drought, allowing them to adapt better to the environment (Boullard, 1957, 1979; Pirozynski & Malloch, 1975). There are two main morphological types of arbuscular mycorrhizas (AM), the Arum-type and the Paris-type and a continuum between them. They are responsible of the P plant nutrition. In the Arum type, fungi form intercellular hyphae between the cortical cells and intracellular arbuscules within them. The Paris-type is characterized by extensive intracellular hyphal coils and arbusculate coils in the root cortex. In the Paris-type the intercellular phase of colonization is almost absent. With very few exceptions, members of single plant species form only one type of colonization. It is often accepted that AM morphology is controlled by plant identity.

Alternation of generations

The number of chromosomes in an individual becomes doubled with fertilization between gametes. This diploid stage continues upto the formation of spore mother cells. Meiotic division takes place at this stage and the haploid (n) spores are formed.

The spores are the beginners of the gametophytic or the haploid stage. The spore germinates and prothallus is formed. The antheridia and archegonia develop on the prothallus. The antherozoid and egg fuse and fertilization ensues resulting in a zygote formation.

The zygote is the beginning of the sporophytic stage (2n). The fern exhibits an alternation of haploid (gametophytic) and diploid (sporophytic) generations.

ECONOMIC IMPORTANCE

The species of *Pteris* are of little economic value. They are noxious and toxic weeds, which may cause harm to livestock. However, some species are useful. The plants have been used as a source of potash for soap and glass making industry. Fronds are used as fuel and thatching material. They are also used as medicine and anthelmintic. Plants are used for brewing purposes, as tan or dye and are also used for making paper and board.

Although modern ranchers and wildlife managers often consider bracken ferns a nuisance, they may be utilized by humans in a variety of ways. In fact, they were so valued in the Middle Ages, when they were commonly used as a source of fuel and as a roofing material, that they could be used as a sort of currency. Bracken ferns were also widely used at one time in the production of potash and bleach, and continue to be utilized in some locales as a food source and as bedding for cattle. The rhizomes of the plants are especially useful. Their extracts can be used to dye wool and tan leather, their starch can be used as a type of flour, and they may even possess medicinal qualities, having been utilized both as a treatment for bronchitis and parasitic worm infections

A number of different animals feed on bracken ferns, although they can be toxic if they are consumed in too great a quantity. The most palatable parts of the plant are the young fronds and the rhizome, a subterranean horizontal stem from which an extensive collection of roots and shoots extend. Interestingly, when the fronds of the bracken fern become damaged, they release hydrogen cyanide gas. Some herbivores, such as sheep, are able to detect this substance, ensuring their avoidance of spoiled leaves.

Class	POLYPODIOPSIDA
Order	POLYPODIALES
Subclass	PTERIDINEAE
Family	PTERIDACEAE
Subfamily	PTERIDOIDAE
	PTERIS

QUESTIONS

Q1. Discuss the salient features of ferns.

Q2. Discuss the catapult mechanism of spore release in *Pteris*.

Q3. Discuss the anatomy of fern petiole and sporophyll. Illustrate your answer.

10

CHAPTER

HETEROSPORY

'Heterospory' *sensu lato* has long been one of the most popular topics of review and research in organismal botany. However, with a few exceptions, reviewers have tended to view heterospory primarily as a precursor to the seed habit rather than a profound innovation in itself. Several workers have discussed and defined **heterospory** and some of these definitions have been provided as:

- "...bearing spores of distinctly different types" (Jones, 1987: 411).
- " ...the condition of producing microspores and megaspores" (Weier et al., 1982: 682).
- "...having two types of spores: megaspores and microspores" (Taylor & Taylor, 1992: 840).
- "...the condition in embryophytic plants in which spores are of two types: microspores and megaspores"(Traverse, 1988: 503).
- "...the spores are of different sizes... "(Sporne, 1975: 11).
- "...two sizes of spore are formed... "(Thomas & Spicer, 1987: 95).
- "...production of microspores that grow into male gametophytes and megaspores that develop into female gametophytes; the two kinds of spores may or may not differ in size"(Bold et al., 1987: 833).

Though the phenomenon has remained of interest to researchers in evolutionary biology and phylogeny besides developmental biology, however, the studies have placed less emphasis on the homospory-heterospory transition than on the heterospory-seed habit transition. Moreover, events like pollination which lead to seed formation are not elaborated. It is a great investment in terms of energy by the plant to organize the female reproductive parts to receive pollen. The chapter throws light on the homospory-heterospory-seed habit transition sequence and emphasizes that a progressive increase in reproductive sophistication occurred via this transition. Heterospory which eventually arose independently in as many as 11 separate vascular plant lineages, includes members of Zosterophyllopsids, Lycophytes, Equisetophytes, the water ferns, and several progymnosperm lineages, amongst them are the ancestors of the extant seed plants (Bateman & DiMichele, 1994). Many of the forms had **incipient heterospory** and several others were free-sporing hetersporous species (e.g., *Selaginella*) and these were the plants which gave way to seed habit as seen in present day plants. This transition in present day seed bearing gymno- and angiosperms was marked by several modifications, the foremost was integumentation.

Stewart and Rothwell (1993) proposed five stages in the evolution of heterospory from the primitive homosporous condition:

1. Decrease in number of spores in some sporangia of homosporous member.
2. Increase in size of the remaining spores in those sporangia.
3. Spore content in those sporangia remained constant in size and number.
4. Change from monoecious to dioecious gametophytes.
5. Change from exosporic to endosporic gametophyte.

In general, heterospory is significantly characterized by two basic features – first the gametophytes are strictly unisexual and thus incapable of self-fertilization. Second - in almost all cases, heterospory co-occurs with endospory, the sexual maturation of gametophytes while housed within the spore wall.

HOMOSPORY-HETEROSPORY TRANSITION

In practice, **homospory** has been observed in most species whether living or fossil, merely on the evidence of unimodal size distribution of small (< 200 //m) spores produced by the sporophyte. The spores are commonly referred to as isospores and the species are said to be homosporous. Such spores upon germination, result in bisexual (monoecious) gametophytes. However, morphologically identical isospores can occasionally generate sexually dimorphic prothalli (monoicy). For example, several extant species of *Equisetum* subgenus *Hippochaete** which are morphologically isosporous but routinely produce gametophytes of two physiological types - some yield only antheridial prothalli (dioiecy), whereas others yield larger prothalli that are initially archegonial but later undergo a transition under some environmental cues to an antheridial condition (sequential monoicy). Such observations have been confirmed *in vitro* by altering environmental parameters specifically (by decreasing incident light intensity and/or prothallial density on the artificial substrate) and obtaining higher ratios of dioecious to sequentially monoecious prothalli in *E. tehnateia**. In this case, no correlation between the sexual expression and spore size was observed. Similar cases of mixed male dioiecy and sequential monoicy have been recorded in a large proportion of the few pteropsids. Like *Equisetum**, *Ceratopteris thalictroides,* forms unimodal spores with no apparent size segregation among sporangia (i.e. the plant is homosporous). It produces a mixture of antheridial and transitional archegonial-antheridial prothalli. Unlike *Equisetum**, there is a strong positive correlation between spore size and gametophyte gender. Smaller spores tend to produce strictly antheridial prothalli, implying gender control via metabolic microenvironments. As one moves a step ahead, similar gametophyte dimorphism is evident in the filicalean pteropsid *Platyzoma microphylla**. In this plant dimorphism is reflected in subtle spore dimorphism and size bimodality that is expressed among sporangia (rather than within). Some sporangia contain ca. 32 spores that consistently generate exclusively antheridial gametophytes, other contains ca.16 larger spores that consistently generate sequentially monoicous gametophytes. In the case of *Platyzoma**, dioiecy is detectable by heterospory and it is therefore, frequently proposed as an evolutionary intermediate between homospory and free living heterospory with endosporic gametophyte. It requires however, moist conditions for successful syngamy.

Homospory, the prevalent ancestral condition in vascular plants has certain ecological and evolutionary limitations. The main disadvantage of homospory seems to be the need for ecological coordination between the gametophyte and sporophyte generations. In particular, the sporophyte must grow at or near the point of syngamy (on the gametophyte with archegonia), a location dictated by the gametophyte. In heterosporous plants, the obligately unisexual male and female gametophytes are functionally analogous to sperm and egg. They are released directly into the environment shortly after meiosis. There is, then, a near simultaneity of spore, gametophyte, and gamete production. One of the key attributes of the extreme compression of the gametophytic phase of the life cycle thus is apparent: gametophytes no longer directly control the sex ratio, as they do in homosporous plants. A functional dichotomy is observed in homosporous plants between sporophytes as the agent of dispersal (via spores) and the gametophyte as the point of syngamy. In heterosporous plants, the sporophyte dictates when, where and under what physical conditions, the gametophytes will perform their functions. Due to lags between the perceiving of environmental cues and the time of release of functional sperm and egg, the potential for the gametophytes from such heterosporous species to encounter unfavourably dry environment in which syngamy cannot occur, is high. Thus, the free-sporing heterosporous plants are far more limited ecologically by the peculiarities of their life history than are homosporous species. How is the sexual reproduction in such situation accomplished? The sporophytes that are tolerant of very dry physically stressful conditions, might not produce gametophytes that are able to reproduce sexually under such dry conditions. The fitness of the dry-tolerant sporophyte in such cases is, therefore compromized. Such plants rely to a great extent on asexual reproduction.

PREVALENCE OF HETEROSPORY

Lycopsida

Extant Selaginellalean members are generally regarded as retaining a single tetrad of viable spores per megasporangium, though some species may show more than one tetrad at maturity in some sporangia (over 20% in *S. lobbii*), whereas in other species apparently abortions within tetrads during development typically leave only one (*S. zvilldenozvii*) or two (*S. lobbii, S. erythropus*) viable megaspores per megasporangium. The order Selaginellales includes species that possess several specialized mechanisms for active dispersal of microspores and megaspores. Megagametophytes are largely endosporic, relying on food reserves traceable to the parental sporophyte, though archegonia and/or rhizoids often project through the triradiate suture. Megaspore wall ornamentation in general and laesural ornamentation in particular often allows the entrapment and transport of microspores produced by the same sporophyte, thereby ensuring fertilization but at the expense of probable autogamy. Once fertilized by a biflagellate spermatozoid, the embryonic sporophyte develops rapidly. The extant *Isoetes** is generally regarded as a descendant of Carboniferous tree lycopsids and most species of *Isoetes,* like their presumptive progenitors such as *Chaloneria,* are strongly heterosporous. In this case, megasporangia and microsporangia are usually segregated both spatially and temporally; the microsporangia mature later in the season, encouraging allogamy. However, at least two extant species of *Isoetes* show a type of anisospory where megaspores and microspores develop in the same sporangium.

Sphenopsida

Several calamitacean cones from the Upper Carboniferous of Euramerica show evidence of low-grade heterospory (Good, 1975): examples include *Calamostachys americana, C. casheoua, C. thomsonnii, Paracalamostachys* (? = *Calamostachys*) *spadiciformis*, and *Palaeostachya andrewsii.* The differentiation between mega- and micro-spores was not much in these. Elaters absent from megaspores have been reported on microspores of at least three of the species. Megaspores were two to four times the diameter of the microspores, and with greater intrasporangial variation in size. Each sporangiophore was typically seen with only one gender of sporangium and the megasporangia far less common than microsporangia. The distinction was evident in cones themselves as most cones show segregation of basally concentrated megasporangiophores and apically concentrated microsporangiophores, though *Calamostachys thompsonii* and *Paracalamostachys andrewsii* were capable of forming unisexual cones. Considered together, these calamitaceans showed heterosporangy but not monomegaspory; thus, endospory and endomegasporangy are also unlikely to have occurred in these genera.

The one intriguing exception to low level heterospory is the much-discussed *Calamocarpon insignis* (carboniferous). This also had elater-bearing microspores that otherwise broadly resemble the megaspores. However, each megasporangium contained only one large (2-3 mm), elongate viable megaspore, whereas the microspores were unusually small for a calamitacean (30-60 pm). The two genders of sporangium were usually but not invariably borne separately on unisexual cones. Some dispersed megasporangium-megaspore units contained well developed megagametophytes, implying that not only monomegasporangiate but also endomegasporangiate condition prevailed, broadly comparable with lycopsids such as *Lepidodendron.*

Pteropsida

The earliest heterosporous pteropsids, the extinct stauropteridalean 'pre-ferns' and water ferns with extant genera have attracted a lot of attention of evolutionary biologists. A single leaf of *Azolla** generally bears both small sporocarps containing a single megasporangium and large sporocarps containing several microsporangia. Occasional bisporangiate sporocarps also occurred, notably in the Upper Cretaceous *Azinia* and Eocene *Azolla priiiiaeza*. Only one megasporocyte developed in each megasporangium and three of the four meiotic products aborted leaving a single, 300-400 µm in diameter megaspore. Both genders of spore mass were surrounded by a mucilagenous periplasmodium of tapetal origin. In the megasporangium this mucilage became localized into several proximally concentrated massulae delimiting a central cylindrical cavity. These massulae acted as buoyancy aids. These in turn were enclosed by the distal portion of the sporangium, which dehisced along with the megaspore-massular unit. The mucilaginous material is also believed to have provided protection to the female gametophyte. On the other hand, each microsporangium released several spherical massulae bearing both microspores and hook- or anchor-like glochidia. The microspores as a cluster were held together by hardened mucilage (massula). In certain cases, the massulae bear trichomes that facilitate coupling to the hairs of the megasporangia. In *Salvinia**, both genders of sporocarp are of equal size and several sporocarps are borne on each modified leaf; the most proximal contains several large megasporangia, whereas the remainder contain many smaller microsporangia borne on a repeatedly dichotomous framework. Each megasporangium contains only one viable megaspore, typically the surviving product of eight megasporocytes, surrounded by a thick cellular perispore that bears a remarkable resemblance to the integument of a pteridospermalean ovule. Each microsporangium encloses a single massula containing the products of 8-16 microsporocytes. The microspores produce spermatozoids while still enclosed in the sporangium; similarly, the female gametophyte eventually protrudes from the megasporangium but remains enclosed by mucilage during fertilization and the subsequent development of the sporophyte. The three extant genera of the Marsileales, *Marsilea**, *Regnellidium**, and *Pilularia**, like the Salviniales, bear several unisexual sporangia encased in sporocarps.

ENDOSPORY: THE KEY ROLE

The endosporic gametophyte development is the key innovation that permits the evolution of heterospory into seed habit. To understand the origin and evolutionary significance of endospory, we need also to consider the ecological limitations that **endospory** places on gametophyte function. The gametophytes of heterosporous plants are strictly unisexual and the microspores and megaspores function as 'gametes'. When the spores are dispersed, they already house a few celled gametophyte. The way gametophytes respond to environmental variability, including conditions unfavorable for gametogenesis, is to enter diapause (suspended development) immediately after release and before "germination". This is a remarkably limiting factor given that spores are released directly into the environment without the potential to form a photosynthetic, free-living, independent structure and significant development remains to occur after they land on the substratum. Evidences suggest that heterosporous plants have therefore, been ecologically dominant only in environments with ample free moisture, notably aquatic and amphibious habitats. In such habitats spores-as-'gametes' are most likely to encounter conditions favorable for fulfilling their functional requirements. Furthermore, because of ecological constraints on successful syngamy, early endosporic plants may have been able to survive only under a very narrow range of ecological conditions. Almost all heterosporous plants determine gametophyte gender epigenetically rather than through sex chromosomes, and fix that gender prior to spore release thus, minimizing the ability of the reproductive phase to respond to environmental vagaries. As in homosporous systems, the sporophyte influences gender by regulating the metabolic microenvironment of the developing gametophyte. Spores ultimately destined to be female are usually produced in metabolically favourable positions on the sporophyte relative to positions of their counterparts. From an ecological viewpoint, endosporic gametophytes effectively function as 'gametes' rather than as a distinct alternative life-history phase. Because they have no exosporic existence, endosporic gametophytes fix little carbon and do not acquire nutrients for continued growth. Consequently, free sporing heterospory suffers from several serious constraints. For example, in many fern species the gametophytes rely on chemical signals (antheridiogens) to mediate population-level sex ratios. Such controls also exist for ensuing gametogenesis and syngamy, and in most species at least some gametophytes retain their potential for self-fertilization. Thus, the gametophyte generation is closely attuned to local environmental conditions and can effectively regulate its reproductive functions through such cues. Not only heterospory but a typical terrestrial free-sporing plant, with endosporic megagametophytes with limited food reserves and limited ability to grow independently, would certainly be selected against strongly, even in the most benign environments.

SIGNIFICANCE OF HETEROSPORY

Prevention of inbreeding and promotion of outcrossing is an important evolutionary sequential step related to heterospory and a milestone in organismal biology. Sporne (1962) argued "that monoecious gametophytes [those that produce both egg and sperm] were much more likely to be self-fertilized than cross-fertilized, unless they were submerged in water. But the dioecious prothalli with unisexual gametophytes in a terrestrial environment would be at an even greater disadvantage, for they might never achieve fertilization at all. This is because being unisexual, they are not able to self- fertilize. The antherozoid [sperm] has to be entrusted with the sole responsibility of finding the archegonium [egg-bearing organ], fertilizing the egg and initiating the sporophytic generation. This is where heterospory may have operated to the advantage of plants with dioecious prothalli." The extrasporangial tissue termed the integument became increasingly prominent and elaborated co-opting for pollen capture (lagenostomy). The megasporangium which in higher plants is an equivalent of nucellus remained attached to the sporophyte, allowing *in situ* pollination and/or /*in situ* fertilization. On the male side, the pollen tubes, initially formed as haustoria for microgametophyte nutrition, subsequently co-opted to deliver spermatozoids (antherozoids) to the megagametophyte (siphonogamy), thus reaching a level of reproductive sophistication seen in, for example, extant *Pinus*.

LIMITATIONS

The ecological considerations of DiMichele *et al.*,(1989) regarding the success of heterospory, suggest that as independent phases, the sporophyte and gametophyte remain under separate selective regimes. The sporophyte determines gender, yet is distant from the timing of gametogenesis so that gametophytes cannot control their sex ratio in response to local population structure and environmental conditions. This life history offers no mechanisms to escape sporophytic hegemony as evidenced in the early vascular plants. This represents an evolutionary regression from the more advanced of the homosporous life histories, wherein gametophytes have regained a measure of control over their own fates.

SEED HABIT: TOTAL SPOROPHYTIC CONTROL

Seed habit is the most complex and diverse means of sexual reproduction in vascular plants. Since their first appearance in late Devonian, seed plants have come to dominate almost every terrestrial ecosystem. Evolution of seed habit consists of at least two independent problems - the delivery of male gametophyte to a female gametophyte that is fixed on parent sporophyte and the evolution of ovular, particularly integumentary morphology.

Technically speaking, seed-producing plants are heterosporous in both phylogenetic and functional terms. Consequently, heterosporous reproduction *sensu lato* can be described as the dominant mode of reproduction in most extant plant communities. Several evolutionary biologists have recognized two main suites of evolutionary innovations leading to seed habit. The first suite concerns modifications to the megaspore and megasporangium. The megaspore abortion resulting in lesser number of prospective female prothalli was a big step forward. The retention on the sporophyte, inside the megasporangium followed by integumentation, and improved sporophytic provisioning of the megaspore was not only a significant adaptation, but confirmation towards seed habit to stay. It was reflected in modified megaspore and megasporangium wall structure and thickness. The earliest ovules in late Devonian, conform to the basic morphology of Archaeosperma with radiospermic symmetry, mostly unfused, converging integumentary lobes and enclosing cupule. The second suite of characters concerns pollination biology. The modifications of the megasporangial unit for microspore capture, delay of resource commitment until pollination and/or fertilization has occurred, pollen tube formation (initially haustorial but later co-opted for siphonogamy), and exclusion of pathogens following pollination facilitated seed habit transition.

QUESTIONS

Q1. Taking example of *Equisetum* and *Selaginella,* discuss the evolution of seed habit.

Q2. Elaborate the significance of heterospory in evolution of seed habit.

Q3. Why should we consider *Platyzoma* sp. an important link between homospory and heterospory?

Q4. Endospory occupies an important place in evolution of seed habit. Discuss.

11 CHAPTER

STELAR EVOLUTION

The combination of the vascular tissues of stems and roots with any other associated fundamental or ground tissue, such as pith and interfascicular regions, is defined the 'stele' or the 'central cylinder' (Esau, 1977). Beck et al. (1982) listed several types of recognized steles and classified them into three basic types: (1) protostele presenting a solid column of vascular tissue (2) siphonostele characterized by a cylinder of vascular tissue with pith in the center and (3) eustele showing separated strands of vascular tissue, usually arranged as a discontinuous cylinder.

Because of the paramount importance of conducting tissues in vascular plants, the stele is one of the oldest concepts to generate interest in plant biology. Term 'stele' was coined by van Tiegham (1886) to include not only the primary vascular tissues, but also the conjunctive tissues – pericycle, vascular rays and the pith if it is present. With this concept van Tiegham and Duliot (1886) put forward the stelar theory. The most significant feature of the theory is that shoot and root are interpreted to be fundamentally similar in construction. Both the organs have a central stele with cortex around it and epidermis as the outermost layer. They categorized stele into three basic types.

The concept of 'Stelar Evolution' has influenced the modern concepts of morphology and progression of architectural configurations of primary vascular systems.

According to the stelar theory, the most primitive stele from ontogenetic and phylogenetic point of view is *protostele,* characterized by the absence of central column of pith. It is the basic type and all other types have been derived from it in course of evolution. In its organization, protostele is a central strand of primary xylem enclosed by phloem. This type also known as *haplostele,* represents the stele of primitive psilophytes such as *Horneophyton* and *Rhynia,* and is also seen in stems of young sporophytes or sporelings of ferns where it later gives way to a cylinder with pith (siphonostele) in the mature plant (Fig. 11.1).

Certain variations occur depending on the contour of the xylem in protostele – core of xylem may be lobed or star shaped, and stele is designated as actinostele e.g., *Psilotum* and various species of *Lycopodium*. In *Lycopodium volubile*, xylem is sponge-like, appearing to consist of separate parallel plates of tissue between and around which occurs phloem, this type is plectostele. In *L. cernuum*, the masses of xylem and phloem are more uniformly distributed, the former as irregular scattered groups embedded in a ground mass of the latter. Such types are mixed protostele (Fig. 11.1).

A stele which is phylogenetically more advanced than protostele has a central column of pith and is termed siphonostele, occasionally also referred to as medullated protostele. Such steles are characteristics of members of Pteropsida. The origin of pith in this stele has been debated upon. According to Jeffrey, pith is always extrastelar in origin and is 'included' in core of xylem. The other view explains pith to be stelar in origin and represent degenerated tracheary elements. Both the views were initially supported by various groups of workers and siphonostele was interpreted to arise either by gradual inclusion of areas of cortical parenchyma or by the phylogenetic reduction of the original tracheids to parenchyma. However later, it was Eames who precipitated the entire controversy by summing up that 'the pith of lycopsid forms of more primitive ferns is without doubt intrastelar in origin and that of higher ferns and other more evolved groups is probably extrastelar'.

Structure of siphonostele in stem shows a greater variation than seen in protostele. It is of two basic types – ectophloic and amphiphloic. In former case, the stem stele consists of only an external cylinder of phloem enclosing xylem, condition designated as ectophloic e.g., *Osmunda* and *Equisetum*. Amphiphloic stele exemplified by *Adiantum* and *Marsilea* has the primary xylem bordered internally as well as externally by phloem tissue, with two endodermal cylinders – one separating cortex from the external phloem, the other is situated between the internal phloem and the pith.

In simplest siphonostele, as seen in plants with smaller leaves (microphyllous), the stele is not much affected and there are no leaf gaps. Leaf gaps are parenchymatous areas in the stele which correspond to the vascular traces departing to the leaves or branches. In plants like *Selaginella* with microphylls, the continuity of the main stele remains undisturbed. This is because the leaf gaps are relatively smaller and successive leaf gaps in the stele do not overlap in the internode because the gaps are scattered. Such a siphonostele without leaf gaps is termed cladosiphonic (Box 11.1). In higher ferns where the shoot axis is short and the leaves are large (mega/macrophyllous) and in close phyllotaxy, the leaf gaps are also prominent. The leaf gaps overlap and the stele in such cases gets dissected in a manner that its continuity gets disturbed, the condition is designated as **solenostele***. If a condition persists where the inserted leaves are so close to each other that in any transverse section the stele shows more than one leaf gap, stele gets broken into a ring of

*Solenostele is support for the theory that siphonostele has arisen from invasion of primary xylem by extra fascicular tissues such as endodermis and pericycle. This is indicative of the extrastelar origin of pith. The interstelar origin of pith emphasizes that innermost vascular tissue or xylem got transformed into medulla. Occurrence of mixed pith is reported in primitive living ferns e.g., *Osmunda* where tracheids lie scattered throughout the pith. Such stele with mixed pith is considered an intermediate between a true protostele and a true siphonostele.

scattered strands. Thus, the most evolved steles as seen in ferns and higher plants are encountered as a ring or a scattered series of vascular bundles. Such a stele is dictyostele e.g., *Pteris*, and the scattered strands get interconnected forming a tubular network where each strand has a central strip of xylem surrounded by phloem (Fig. 11.1). These concentric strands of vascular tissues are called meristeles. Each meristele has a basic structure of a protostele. In certain dictyosteles, the leaf gaps may not be the only perforations in the stele, there may be gaps other than leaf gaps and such is known as perforated dictyostele. In ferns like *Pteridium aquilinum*, a number of separate steles arranged in more or less regular cycles one within another develop. This is polycycly. Such steles are always siphonostelic. Polycycly is different from polystely where two or more parallel steles lie side by side in the stem as seen in *Selaginella* (Fig. 11.1).

Going back to siphonostele, in cases where stelar evolution has proceeded to highest evolution, a third type of stele, **eustele** is formed due to dissection of ectophloeic siphonostele. It is frequently met with in angiosperms, the dicotyledons where the interconnected strands are called collateral bundles. If the stele of an angiosperm stem with internal phloem is conspicuously dissected, the individual vascular strand is frequently bicollateral with median xylem strip flanked by phloem on either side.

The most complex stele is the atactostele, as in monocotyledons. It may be interpreted as a monostele where a number of scattered vascular strands are embedded in a conjunctive ground tissue.

Box 11.1

In all the siphonosteles, conspicuous strips of parenchyma known as gaps develop at various points in the vascular cylinder. This is seen in Pteropsida and higher plants as leaf gaps. These are the areas occurring above the point of divergence of vascular strands or leaf traces that occur below the leaves. In highly evolved systems, these leaf gaps, the parenchymatous areas may vary in number and vertical extent. They tend to overlap within each internode and may or may not have any relationship to the leaf trace system. In contrast to macrophyllous plants as ferns, the microphyllous plants such as Psilopsida, Lycopsida and Equisetopsida are devoid of such structures. Siphonosteles devoid of true leaf gaps (such as presence of Branch gaps) are known as cladosiphonic in contrast to phyllosiphonic condition as in Pteropsida.

LIMITATIONS

According to the theory, stele is the real entity and essential part of root and stem. Though the theory got wide acceptance from scientists all over, yet the theory could not fully explain the architecture of modern day plants. The major drawback of this concept is that it does not explain with clarity the vascular cylinder in higher vascular plants like ferns, gymno- and angiosperms. In such plants where the leaf traces are large in number and appear to play a major role in structural organization of axial vascular system of the stem, the vascular cylinder would be a composite structure. Hence, how much of the vascular system of stem belongs to the leaf and how much of it to the stem proper is not clear. For example in microphyllous plants like *Selaginella* and *Lycopodium,* leaf traces form only a very small proportion of the stem vascular cylinder which is not disturbed by the gaps while in megaphyllous or macrophyllous plants, as in ferns, the leaf traces are prominent, more in number and greatly influence the stem vasculature. Wetmore & Wardlaw (1951) regarded the axial system in Filicophyta, therefore as a composite structure containing both cauline (of stem) and foliar (of leaf) origins, with latter varying in contribution.

Another controversial aspect of the theory remains the anatomical separation of cortex from stele. According to van Tiegham, the inner boundary of the cortex is represented by endodermis, characterized by casparian strips chemically composed of lignin and suberin. This type of interpretation however, is acceptable for only lower plants while in seed plants, the boundary between stele and cortex is difficult to interpret. Similarly the interpretation of pericycle for lower vascular plants is that it is a cylinder of cells at the outer edge of the stele of roots and stems, while in higher plants, it represents the outermost portion of primary phloem, or the protophloem that comprises pericycle fibers. According to this interpretation, the independent zone separating cortex from stele is absent (Box 11.2).

Despite the difference in interpretation of tissues in lower vascular plants and higher plants, stelar theory has emerged as a unified concept which has made it possible to classify and trace the phylogeny of various types of steles that occur in stems and roots.

Box 11.2 Endodermis and Pericycle – A quick guide

According to Guttenberg (1943), the term endodermis should only be used for cells with Casparian bands, and not for cells in a ring around vascular tissue that might contain starch. As reviewed by Esau (1953), such a cell layer surrounding vascular tissue and rich in starch would be considered: a "starch sheath" and is "homologous with the endodermis" and may be termed "endodermoid cells. Later, Esau, (1977), redefined the concept and wrote - "Broadly defined, the term endodermis, and endodermoid as an adjective, are applicable to the boundary in any of its manifestations". Soukup & Tylová *(2018)* stated that "leaf bundle sheaths and starch sheaths of the stem are analogous to the endodermis".

According to Beeckman & Smet (2014), the pericycle is a unique layer of cells in plants, named after its position, encircling the vascular tissue in stems and roots. In roots, it is surrounded by the inner cortical layer, namely the endodermis. The pericycle is a heterogeneous, non-vascular tissue in plants that is divided into two populations — one at the xylem pole and one at the phloem pole. Pericycle cells at these poles are marked by differences in size, by ultrastructural features and by specific proteins and gene expression. Transcriptional evidence suggests that pericycle cells are intimately associated with their underlying vascular tissue instead of being a separate concentric uniform clonal layer. Moreover, distinct functions have been attributed to xylem pole versus phloem pole pericycle cells, countering the idea of a delineated plant tissue.

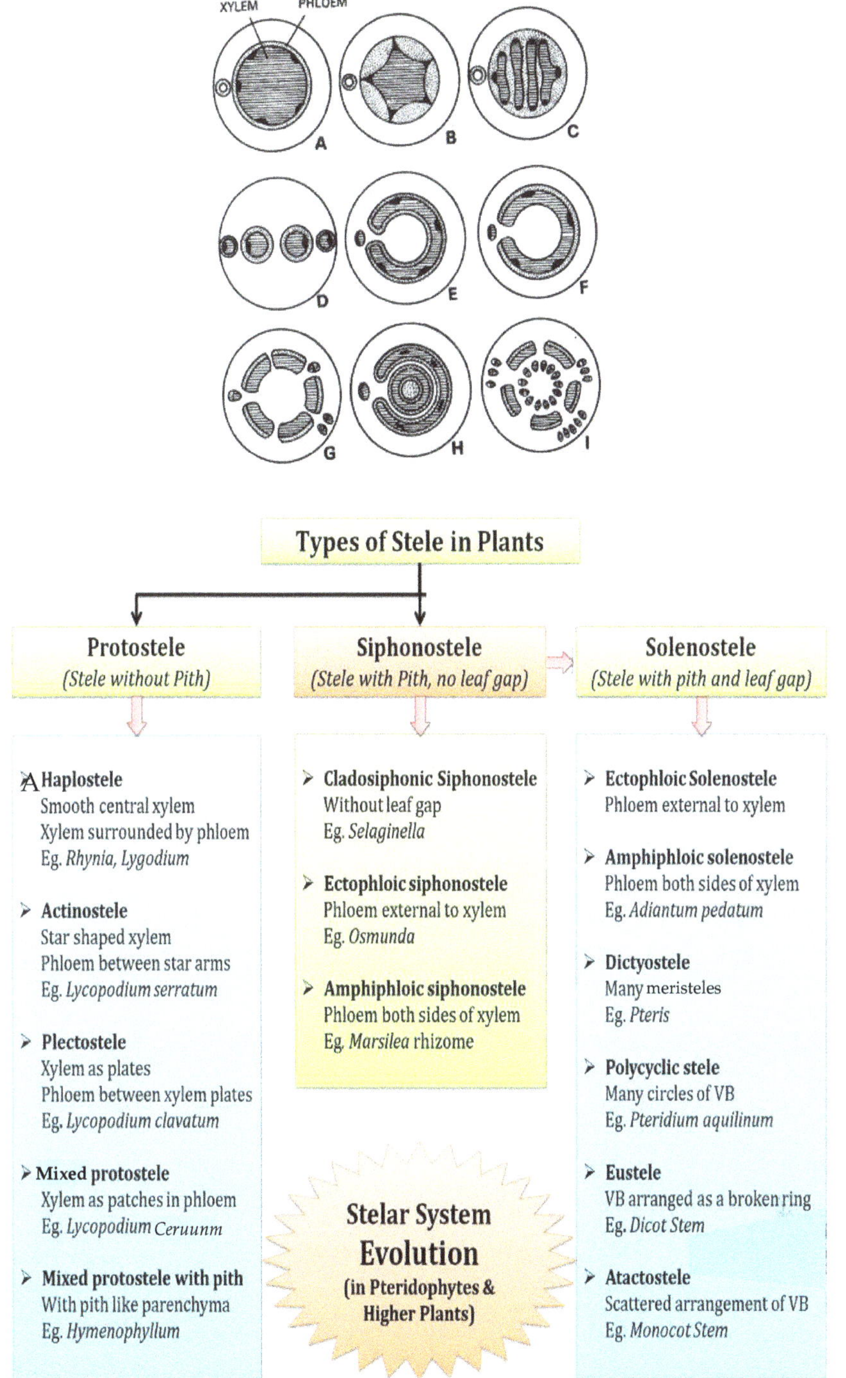

Figure 11.1 A-I: Types of steles in plants

A-Protostele, B-Actinostele, C-Plectostele, D-Polystele, E-Amphiphloic siphonostele, F-Ectophloic siphonostele, G-Dictyostele, H-Polycyclic solenostele, I-Polycyclic dictyostele

QUESTIONS

Q1. Giving examples, discuss the stelar evolution in vascular plants. Illustrate your answer.

Q2. What are the drawbacks of stelar theory?

Q3. Give the salient features of stelar theory.

Q4. Define eustele, meristele, siphonostele, protostele.

Q5. What are the various types of protosteles? Give example of each type.

Q6. Elucidate the evolution of atactostele from protostele.

12

CHAPTER

TELOME THEORY

The most primitive vascular plant perceived by the proponents of the theory was composed of a system of telomes, the ultimate portion of a dichotomizing axis.

Telomes, the standard parts were first observed in fossils, and could be seen in their unaltered form in the simple multicellular stem modules that make up the branching systems of the earliest land plants. However, during evolution, in certain cases, telomes, the ancient structures, may have got modified to an extent that their distinctive identity is no longer recognizable. In its simplest form, telome is terminal portion of the dichotomously branched plant axis while the sections of the plant body connecting telomes are called mesomes. Telomes may be grouped together in various ways to form syntelomes.

The telome could bear a sporangium and be referred to as fertile telome or be a sterile telome when it was referred to as the phylloid. The fertile and sterile telomes may develop in separate or combined groups called telome trusses (Fig. 12.1A,B).

Thus, telome is the basic unit of plant structure. There is justification regarding the fundamental unit of the telome as a real structure, at least in the early land plants of the Devonian Period (408 to 363 million years ago). Early fossils (Fig. 12.1C) had simple dichotomous branching systems but no significant appendages (Gensel & Andrews, 1984; Rothwell, 1995; Kenrick & Crane, 1997a). Each node was composed of a single telomic unit. The only significant variation was the presence of sporangia, and in many species these also had a strong resemblance to stems. In such cases, telomes could therefore, be seen as modified sporangia in which the spore precursors have been diverted to the production of vascular and ground tissue systems. Fossil evidence showed that the degree of branching and its complexity increased with time, and the gradual specialisation of lateral branches to produce organs recognizable as leaves and cones appeared to proceed through intermediate stages where telomic units were still clearly visible.

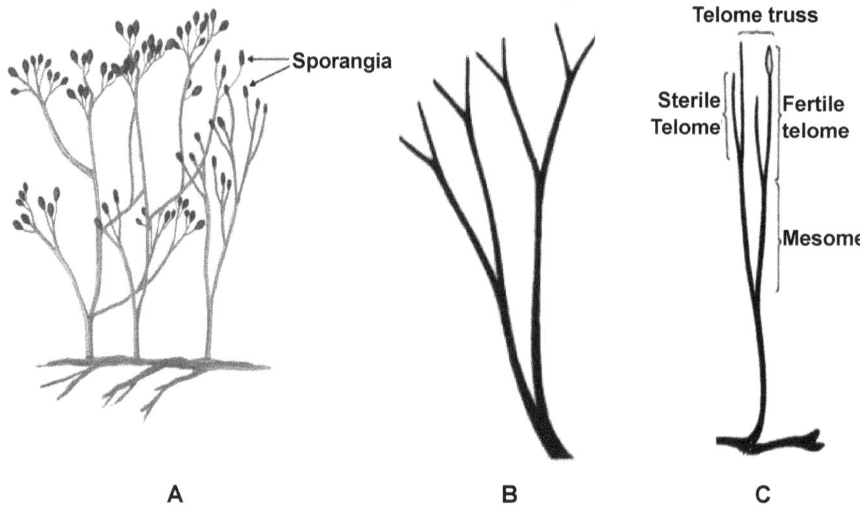

Figure 12.1A-C: (Concept of telome)
Fertile telome (A), sterile telome (B) and a fossil with simple dichotomously branched plant body made up of telomes (C).

CONCEPT AND THEORY

In land plants, telome is essentially a fundamental supercellular building block: an intermediate category between cell and 'fully-fledged' appendage or organ system (i.e. leaf, cone, complex stele). Conceptually, it is a cylindrical axis that develops from an apical meristem. Anatomically, it is simple, comprising a single vascular cylinder (protostele), a parenchymatous cortex, and an epidermis. The concept of the telome was modelled on discoveries of early fossil plants from the Devonian Period, in particular the famous early land plants of the Rhynie Chert (Kidston & Lang, 1921). Telomes can be seen in their unmodified form in plant fossils such as *Cooksonia* (Fig. 12.2) and *Rhynia,* built from a simple set of cylindrical elements that were repeated to give a dichotomously branching system. These examples of simple stick-like plants seemed to provide a basic framework from which many of the familiar organs and appendages of land plants could be derived. Other more complex Devonian plants also appear to be built on a similar body plan, and the fossil record documents a series of morphological intermediates between branching systems constructed from numerous telomes and certain types of plant organ (e.g. leaf, seed). Based on palaeobotanical insights such as these, Zimmermann originator of the theory proposed that land plant morphology was reducible to the 'atomic unit' of the telome. Present day plants are built from numerous simple standard parts that have been combined and modified in various ways to produce the vast collection of organs and appendages seen today. This is the central tenet of the telome theory.

Figure 12.2: *Cooksonia*

Cooksonia- Individuals were small, a few centimetres tall, and had a simple structure. They lacked leaves, flowers and roots, it has been speculated that they grew from a rhizome that has not been preserved. They had a simple stalk that branched dichotomously a few times. Each branch ended in a sporangium or spore-bearing capsule. In his original description of the genus, Lang described the sporangia as flattened, «with terminal sporangia that are short and wide», and in the species *Cooksonia pertoni* «considerably wider than high (Fig.12.2).

Zimmermann, argued that the remarkable diversity of organ systems in plants could be explained by the modification of telomes through the interplay of just a handful of elementary developmental processes. The concept did not arise *de novo*. There were several workers whose understanding and research helped Zimmermann conceptualize the theory. The contribution of Bower (1884) that leaf was a fundamentally flattened branch system, was significant. The beginning was made with a forked marine alga that gave rise to a branched system of early land plants. The leaf and stem resulted from overtopping in this dichotomous plant. According to Bower, the leaf of the ferns and other early vascular plants was dichotomous in organization. He further added that a monopodial pinnate leaf was derived through dichotomy while passing a sympodial stage and the true stem evolved. This flattened leaf was the result of webbing. Another significant contribution was made by Lignier who conceived the sporophyte in primitive land plant to be a dichotomously branched system of cylindrical cauloids (of stem origin) bearing leaf-like appendages (or phylloids) on aerial part and rhizoids on the underground part. Sporophytes which were relatively undifferentiated, were protostelic. The part of this cauloid system which remained underground got transformed into root system. The cauloids bore terminal sporangia, possibly bivalved. Thus, there was no major difference between leaf, stem and root.

Zimmermann used these concepts to formulate telome theory emphasizing that the plant body was a branched axis with a descending segment, the root and an aerial segment, the shoot with merely modified structures. The groups of terminal cauloids however, became dorsiventral in symmetry; the cauloids within each cluster fused together, thus forming a flat plate or leaf blade in which the constituent cauloids were reflected as dichotomous venation. The appearance of megaphyllous leaf brought about the disappearance of the phylloids which became useless. The main tenets of Zimmermann's theory are dealt under three main heads namely, a) the forerunners (ancestors) of primitive land plants, b) the primitive land plant and, c) derivation of the plant body of the higher land plants.

PRINCIPLES OF THE THEORY

a) The Ancestors of Primitive Land Plants

The ancestors were algae, that lived in tidal zone of Cambrian and Silurian seacoasts and were undifferentiated, radially symmetrical and composed of dichotomously forked branches (primitive telomes). Internally such plants possessed a central strand of mechanical tissue (sclerenchyma) and exhibited isomorphic alternation of generations. Zimmermann in this concept advocated therefore homologous theory of alternation of generations of land plants. According to him, land plants inherited basic body plan from their green algal ancestors.

b) Primitive land plants

Zimmermann supposed that the aquatic ancestors of land plants had an isomorphic alternation of generations in which both gametophyte and sporophyte had well-developed telomic branching systems (with telomes and mesomes). The stele in early land plants was simple – a protostele derived from sclerenchymatous mechanical tissue of marine alga. The reproductive structures were terminal sporangia which produced spores, product of meiosis. The sporophyte was dependent on gametophyte and reduced to a single telome. Telome theory is therefore, psilopsid centered for the upper Silurian to mid Devonian Psilophytales (*Rhynia, Horneophyton*) as the exemplified sporophytes of primitive land plants.

It was envisaged that the gamete- and spore-bearing phases of the life cycle - the gametophyte and sporophyte, underwent different fates following the transition to the land. In vascular plants, the gamete-producing phase was reduced whereas the spore-producing phase was elaborated. Both phases underwent some loss of morphology, but this was most marked in the sporophyte, which was reduced to a single telome and became parasitic on the gametophyte (in bryophytes, life cycle evolution followed a different path). The basic telome morphology was therefore, forged in an aquatic environment, and its origin pre-dated the transition of plant life to land (Fig. 12.3A).

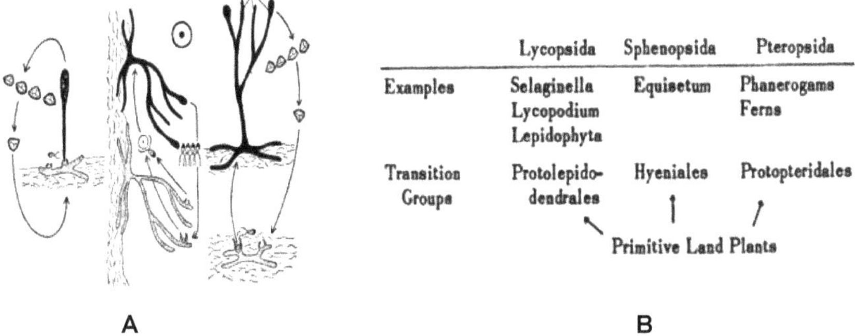

Figure 12.3A,B: Telome concept and evolution of land plants.

c) Derivation of the plant body of higher land plants

The higher land plants evolved from telome trusses of the primitive forms along three parallel lines – Protolepidodendrales, Hyeniales and Protopteridaes (Fig. 12.3B) and later evolved through morphological specializations called elementary processes (Table 12.1).

Table 12.1: Elementary processes in the formation and modification of telomes (Zimmermann, 1952)

Process		Result
Telome origin		
1.	Connection of cells through common cell walls	Filamentous growth
2.	Rotation of plane of cell division	Branched filaments
3.	Apical meristem	Cylinderical multicellular body
4.	Changes to isomorphic life cycle (ancestral isomorphic life cycle an assumption of the theory)	Elaboration of either sporophytic (vascular planes) or gametophytic (bryophytes) phases of life cycle at expense of other
5.	Tissue differentiation	Elaboration of ground tissue system and vascular system
Telome modification		
6.	Overtopping	Stronger development of one branch in a dichotomy-distinction between main axis and lateral appendages
7.	Planation	Switch from multi-dimensional to flattened branching
8.	Syngenesis	Lateral fusion of telomes– implicated in webbing of leaves and development of complex vascular systems

	Process	Result
9.	Reduction	Loss of telomic units leading to loss of morphological complexity (e.g. bryophyte sporophyte)
10.	Incurvation	Bending due to unequal growth of tissue on flanks of telome
11.	Longitudinal differentiation	Development of serial homologues along an axis (e.g. leaves or pinnules)

ELEMENTARY PROCESSES

If telomes are the fundamental building blocks of plants, then how did they evolve, and what processes have sculpted them into the leaves, stems, seeds and so forth of modern plants? Zimmermann proposed about a dozen specific and independent 'elementary, processes' that gave rise to and operated on telomes. He argued that all the major organs and organ systems of plants are the outcome of elementary processes acting on their own or more usually, in combination. By applying these processes to the simple branching systems of early land plants it is possible to transform them into more complex structures'.

Overtopping– Two sister branches of a telome truss (equated to syntelome) which were initially equal, became unequal and the stronger became vertical while the weaker was pushed aside, became overtopped. Thus, from equal dichotomy to unequal dichotomy to a sympodial and finally monopodial system. The lateral members then formed leaves. In a pinnately compound leaf, overtopping mesomes formed the rachis and midvein, and the overtopped mesomes constituted the leaflets and lateral veins.

Planation– Branching in more than one plane is replaced by fan-shaped dichotomy bringing all telomes and mesomes into a single plane. Thus, telomes and mesomes arranged themselves in a plane. The radial symmetry gave rise to bilateral symmetry. This stage explains leaf formation (Fig. 12.4A,B).

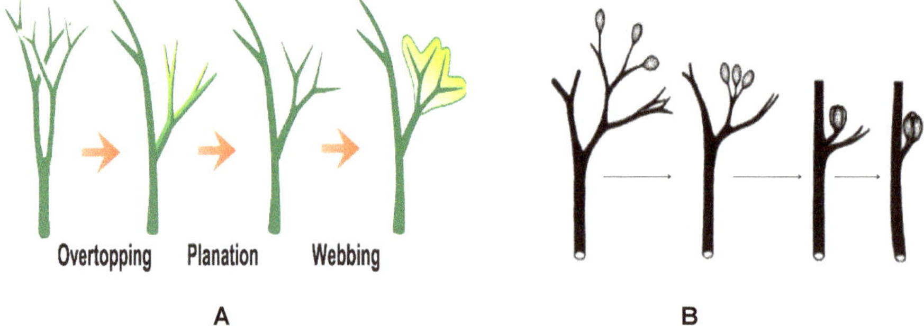

Figure 12.4A,B: Elementary processes in organization of plant body – (leaf in A and sporangium in B) in higher plants with telome as a fundamental unit.

Webbing or fusion– Is connection between telomes and mesomes by formation of parenchyma between them. In simplest case the phylloids became united by development of parenchyma between them thus forming a foliar appendage with open dichotomous venation. Similarly in shoot axis, a purely parenchymatous webbing led to polystelic condition as seen in *Selaginella.* It explains formation of both, the leaf and the stem through the degree of webbing.

Reduction– Simplification of telomes truss leading to formation of a single needle-like leaf.

Origin of leaves has also been explained on the basis of Telome theory. Foliage leaves are perceived comparable to the last branching of a telome truss. They develop in two directions to form microphyll and megasporophyll. If each phylloid forms directly, a flattened single nerved needle as in *Lycopodium* and *Equisetum,* a microphyll is formed, but if all the phylloids of a truss together form a large bladed leaf as in ferns and angiosperms, this is megaphyllous type (Fig. 12.5A,B).

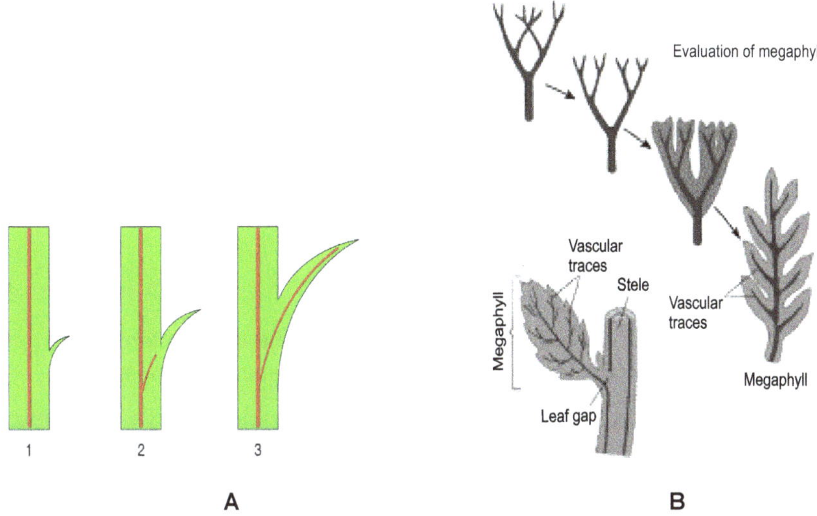

Figure 12.5A,B: Origin of leaf enations of lycopods (A) and megaphyllous leaves (B).

Thus, microphylls and megaphylls are of similar origin, both differing in number of phylloids involved. The large complex leaves of ferns represent major branch system that has flattened in formation. The leaves of Arthrophyta (*Equisetum*) represent a minor branch system, and the number of telomes entering its formation is smaller. However for Lycophyta, leaves have been called as enations – new structures that have appeared as lateral outgrowths from a surface previously unoccupied. These outgrowths or spine-like structures appeared on naked stems, evolving into longer flattened leaf-like appendages with time. These appendages were first without any vascular supply then with a leaf trace that terminated at the base of appendage and finally a vein or vascular supply that ran from base to the tip (Fig. 12.5A,B). In this group of plants therefore, microphylls evolved by a gradual process of enlargement, rather than by progressive reduction from a larger structure. Evidence can be traced from Psilophytales with *Psilophyton* possessing a small scale-like leaf without any leaf trace, followed by *Asteroxylon* with a rudimentary leaf trace and *Drepanophycus* with a leaf trace going into lateral appendage giving way to *Tmesipteris* which has foliar organ with a well-defined vascular strand (Fig. 12.6A-D).

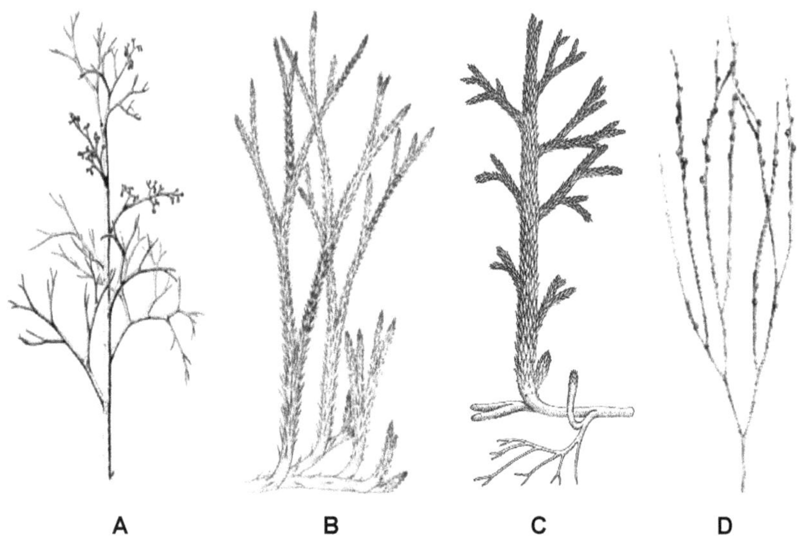

Figure 12.6A-D: (Fossil plants with telome as the unit of plant body) A- *Psilophyton*, B- *Asteroxylon*, C- *Drepanophycus*, D- *Tmesipteris*.

Fertile telomes have also become arranged in various ways by elementary processes. Such processes are aggregation, fusion, overtopping, reduction recurvation and incurvation. According to telome theory, sporophyll is composed of one or more telomes or both telomes and mesomes. An aggregation of fertile and sterile telomes followed by reduction of supporting mesomes and reduction of sporangium number leads to one phylloid and one sporangium. Further reduction leads to fusion of sporangium with shoot axis, thus becoming cauline in position as seen in *Lycopodium.* The sporophyll of *Sphenopsida* is believed to have occurred due to downward bending of sporangium towards the main axis by recurvation and finally fusion and expansion of telomes and mesomes into peltate discs. The sporophylls of Pteropsida are visualized formed by overtopping, fusion and reduction which lead finally to sporangia along the margins of the sporophyll. Incurvation or unequal growth on the two opposite flanks of an organ often leads to shifting of the sporangia on the lower surface of the sporophyll (Fig. 12.7A- P). Further incurvation leads to simple carpel and fusion of carpels gives rise to various ovaries.

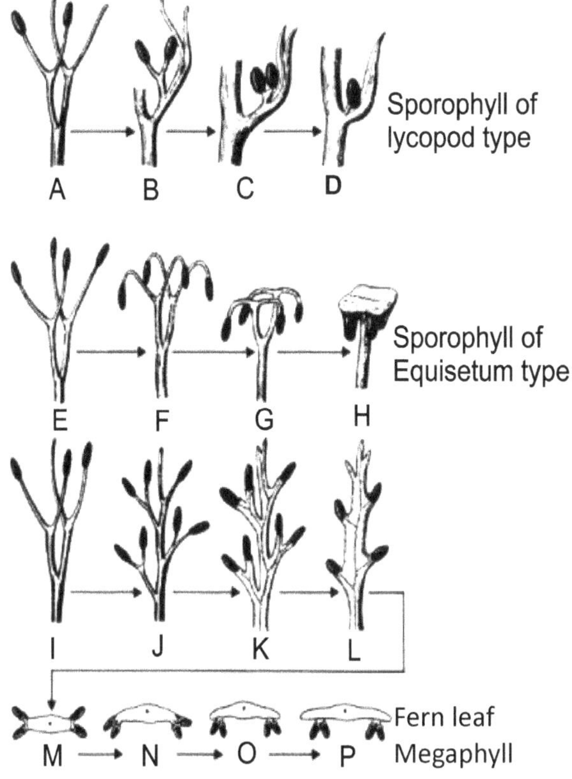

Figure 12.7A-P: Stages in formation of fertile structures.

Co-option and modification of sporangia to form leaves and ligules in lycopods

Lycopods are an ancient group of vascular plants with a very distinctive and highly conservative morphology (Jermy, 1990a, b; Bllgaard, 1990; Stewart & Rothwell, 1993). Small herbaceous species of the Devonian period strongly resemble their modern relatives. So similar are they that *Asteroxylon* from the early Devonian, is presumed pre-lycopsid. There are three plausible hypotheses for the origin of the lycopod leaf (Crane & Kenrick, 1997):

(1) **Reduction hypothesis–** Zimmermann (1959) suggested that the leaf was a reduced telomic branching system. He based this hypothesis on the observation that many early land plants have lateral branching systems. Furthermore, one extinct group of early lycopods had forked leaves, which could be interpreted as vestigial evidence of branching.

(2) **Enation hypothesis–** Bower (1908) suggested that this leaf type evolved as a completely new outgrowth of the stem. He pointed to evidence of simple,

spine-like non-vascular appendages in many early land plants, which could be interpreted as intermediate stages in the evolution of a fully vascularized leaf (Fig. 12.8A,B).

(3) **Sterilization hypothesis–** Kenrick & Crane (1997a) suggested that the leaf evolved initially as a bract to the sporangium through duplication and sterilisation of one sporangium in a pair. This bract was later co-opted to a photosynthetic function, producing a leaf. They emphasised ontogenetic, positional and anatomical similarities between sporangia and leaves and pointed to evidence of forms with sporangial-shaped bracts in the early fossil record. The ligule was also explained as a further duplication of the leaf pathway. Both the reduction hypothesis and the sterilization hypothesis are telomic in nature because of their emphasis on co-option and modification of lateral branches or sporangia.

BODY PLANS AND PHYLOGENY

Within vascular plants, Zimmermann recognised three major body plans that corresponded to three monophyletic groups: lycopods (clubmosses), sphenopsids (horsetails) and pteropsids (seed plants, together with ferns); and the fourth one in non-vascular - bryophytes. The body plans of these three groups evolved from within a grade of extinct plants with telomic branching modelled on the early fossils *Rhynia* (now *Aglaophyton)* (Edwards, 1986; Remy & Hass, 1996) and *Psilophyton* (Banks et al., 1975). The telome theory, therefore makes important assumptions, or perhaps statements, about phylogenetic relationships.

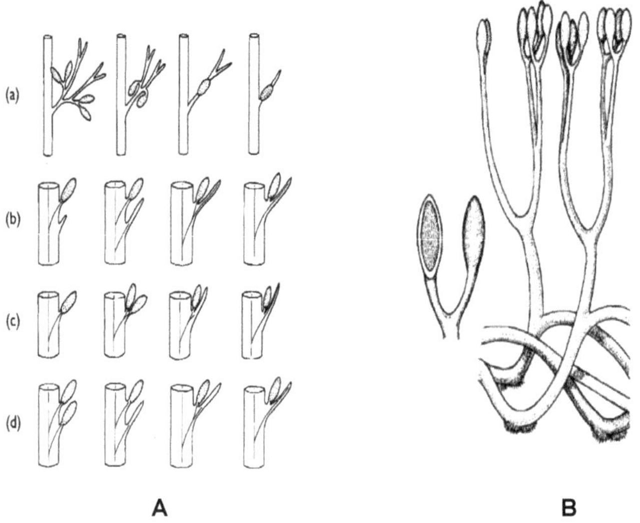

Figure 12.8A,B: Formation of a ligule in lycopods (A).
B- *Aglaophyton* – a hypothetical fossil.

EXPLANATORY VALUE

The theory emphasizes the concept of homology and is fundamental to research in comparative biology even in present times. Telomes are standard parts that have been combined, shaped and moulded through time into the full gamut of complex organs that characterise the modern plant. From developmental and evolutionary perspectives, the telome theory is economical because the use of standard parts to make complex organs actually reduces the complexity of genome-level processes required to encode and control development. By using standard parts, a plant can generate a more complex morphology than could be encoded in a one-to-one fashion by the genome, and it reduces the complexity of the command system during ontogeny (Raff, 1996).

SIGNIFICANCE

The theory puts forward the idea that organisms are composed of standard parts such as stem, root and leaves and that these can combine and metamorphose into different organs. Thus all parts of the plant such as leaves, petals, seeds and flowers are not only composites of standard parts but also appear to be related among themselves. The theory has been widely used to explain the origin of the familiar organ systems of ferns, lycopods and gymnosperms. The theory has been able to describe structure of sporophyte in earliest land plants besides discussing morphology of leaf, root and aerial part of the plant

The telome theory attempts to explain the origin and development of complex plant organs from more basic multicellular units called telomes, central to the formulation of theory. These are standard parts that were first observed in fossils (Fig. 12.8B), and they can be seen in their unaltered form in the simple multicellular stem modules that make up the branching systems of earliest land plants. The appeal of the telome theory as an explanation of morphogenesis in higher plants lies in its apparent applicability across a wide variety of organs and taxa.

The telome theory brings together a set of approaches, ideas and assumptions that represent the culmination of a movement that came to be known as 'New Morphology'.

The telome theory states that the organs, appendages and vascular systems of land plants evolved from simple multicellular modules (telomes) that have been combined and modified in various ways. Zimmermann's general phylogenetic outline of plants holds up well. Modern studies support monophyly of land plants and vascular plants (Kenrick & Crane, 1997b; Kenrick, 2000; Pryer et al., 2001; Schneider et al., 2002). Three major clades of extant vascular plants are recognised and these broadly correspond to Zimmermann's three body plans (clubmoss, horsetail, fern+seed plant). There is general agreement that early land plants of the *Rhynia* and *Psilophyton* type are a basal grade of organization within vascular plants.

THE PRESENT INTERPRETATION

With regard to vascular plants, at the time when Zimmermann formulated the theory there was no direct evidence on the life cycles of *Rhynia* and its relatives. They were believed to have isomorphic alternation of generation. However, recent discoveries have shed much light on this concept (Remy et al., 1993; Kenrick, 1994). Remarkable discoveries have been made from Rhynie Chert which show although similar, but the gametophyte and sporophyte not strictly isomorphic, the gametophyte in *Rhynia* is smaller than the sporophyte. The exceptional association of branched sporophytes and a basal structure interpreted as a reduced thalloid gametophyte in *Cooksonia* reveals the existence of heteromorphic, sporophyte dominant, life cycles for the earliest eutracheophytes (Eutracheophyta) (Remy et al., 1993). Unexpectedly, the life cycle in the earliest vascular plants was isomorphic in a number of taxa and in many early vascular plants, the gametophyte was a reduced structure. Unlike living vascular plants, both spore-bearing and gamete-bearing phases had well-developed telomic branching. The telome theory has proven correct in its interpretation that the vascular plant gametophyte has clearly undergone a spectacular loss of morphology and a drastic reduction in size, whereas the sporophyte has been the focus of most subsequent morphological elaborations.

DRAWBACKS OF THE THEORY

Complex vascular systems - the telome theory rejected

Despite the success telome theory received, it never had a major impact on angiosperm morphology. The main problem with the theory is that majority of the workers believe it to be more of a concept than a theory. As argued by present day workers it is a description of terms rather than a theory and has lot of imagination and lacks predictive power. A telomic approach to understanding morphology within ferns, the more derived gymnosperms, and angiosperms is simply inappropriate. The reason for this is that telomes in these groups have metamorphosed into a new and more diverse set of standard parts (e.g. leaves, seeds, sporophylls, petals, stamens). Telomes in such cases are no longer part of the terminology of the morphologist because they no longer exist as discrete, identifiable entities. The theory has made some general predictions and there is no proof for such transitions in the fossils. The concept of the telome does however, remain an important one for understanding the evolution of some fundamental organ systems in plants. Telomes are clearly recognizable as distinctive modules in the simple dichotomous branching systems of the earliest land plants. Zimmermann's ideas on leaf evolution through modification of telomic branching systems also remain valid hypotheses.

One of the most problematic of Zimmermann's many hypotheses of organ transformation relates to the evolution of complex vascular systems in higher plants, many evidences for transitional stages are missing. Telomes have a simple vascular supply comprising a centrally located terete xylem strand surrounded by a zone of phloem. The vasculature in most of the higher plants is, however, much more complex. The range of variation is broad, and when seen in transverse section, encompasses lobing of varying degrees, medullation of the pith, dissection and multiple steles (Ogura, 1972; Beck et al., 1982). Zimmermann's explanation for this diversity was novel. He proposed that complex steles evolved through lateral fusion of multiple telomes (syngenesis, each adding its own vascular system to produce a complex whole). Under this interpretation, the shoots of a large number of higher plants are most definitely compound structures. Syngenesis should lead to a polystelic condition resulting from a fusion of lateral fusion of multiple telomes.

As a general explanation of complex vascular systems, the telome theory has several problems. First, there are no examples of complex vascular systems preceded by transitional forms with false trunks (i.e. trunks composed of partially fused stem elements) in the fossil record. Those plants that do produce false trunks represent comparatively derived groups. The trunks are generally formed from closely growing petioles (e.g. Osmundaceae: Ogura, 1972), roots (Cyatheaceae: Ogura, 1972), or stems bound together in a root mantle (e.g. Tempskya: Andrews & Kern, 1947). These taxa are neither closely related to early land plants with complex steles (e.g. Cladoxylales) nor do they precede them in the fossil record. Second, in plants with complex vascular systems, there is no supporting developmental evidence for fusion of any kind (Stein, 1998). In fact, it is well established through developmental studies that stelar architecture is induced by hormonal gradients set up by shoot apices and lateral appendages (for discussion see Wight, 1987). Third, many of the complex stelar patterns observed in early land plants can be reproduced by models that simulate lateral appendages and their associated hormonal fields (Stein, 1993). There is therefore, a broad coalition of data from systematics, the fossil record, experimental studies of living species, and modelling that is inconsistent with the telome theory. Furthermore, there is a more plausible alternative hypothesis for the origin of complex steles in higher plants (i.e. induction via meristem-mediated hormonal gradients). These would seem to be sufficient grounds for rejecting the telome theory as a plausible hypothesis in the evolution of complex steles.

QUESTIONS

Q1. Telome theory has convincingly explained the body plan in early land plants. Comment.

Q2. Discuss the elementary processes which have resulted in formation of microsporophylls and megasporophylls in early land plants.

ECONOMIC IMPORTANCE AND BIOLOGICAL INTERVENTIONS

In general when we discuss the economic value of pteridophytes, it is the large group of ferns which comes to one's imagination. However, ferns with little economic importance attached, are popular horticultural plants with many species grown in ornamental gardens or indoors. Ferns immediately capture the attention of all who visit moist shady areas in hilly regions or some dense patch of vegetation. With their large, highly dissected and shiny green leaves, ferns are so visually appealing that many species are said to have an aesthetic charm.

Some of the amazing roles that ferns can play:

- Provide microhabitats, as well as shelter and shade to small animals
- Provide a source of food or medicine for animals, and humans
- Ceremonial and spiritual use
- Colonize disturbed sites as one stage in succession
- Remove or bioaccumulate toxins, such as heavy metals, from environments and thus are good bioremediators
- Provide a bioindicator for the health of an ecosystem
- Evolve to fill unique niches in ecosystems and co-evolve with other species (often endemic)
- Ferns provide natural vegetation that is important to both wildlife and birds

ECOLOGICAL ROLE

Most moist woodlands have a number of fern species blanketing the understory with their attractive foliage. In tropical woodlands, ferns are often at eye level or above, and provide an aesthetic and delicate subcanopy. Even in arid lands or on newly exposed surfaces such as burnt soil, or landslides, ferns are present and sometimes are so dominant, that they catch full attention of passer-by. Beyond their immediate visual appeal, ferns are curious objects. Very little is known about how ferns interact with their environment. Certain species are weedy colonizers and when introduced, have displaced native plants in some parts of the world. On the other hand, there are certain species which have either become endangered or rare majorly due to habitat destruction. Successful control measures for invading species, as well as the conservation of rare ferns, are part of many natural ecosystem management strategies and require a thorough understanding of fern ecology. *Pteridium* is an extremely successful colonizer worldwide and can form dense, nearly impenetrable stands. It generally contains several biochemical compounds such as high levels of carcinogens that make it a health hazard when consumed by livestock or humans although there are reports that the early Maori of New Zealand used *Pteridium esculentum* rhizomes as a crop. Where it becomes an invasive weed, *Pteridium* is difficult to eradicate. *Pteridium* therefore continues to be an important subject for ecological research (Robinson, 2007). *Lygodium microphyllum* (small-leaf climbing fern) from the Old World was introduced into gardens in the southeastern USA in the 1950s, but has now become a serious pest in vast areas (Hutchinson & Langeland, 2006). Tu & Ma (2005) recently reported that the hyperaccumulating *Pteris vittata* (Chinese brake fern) can remove quantities of otherwise inaccessible arsenic from mined wastelands. While as a bioremediator, it could be of great benefit, understanding the ecology of *P. vittata* before remediation is now essential because it has been identified as an invasive fern (Palmer, 2003). *Azolla* another fern species is also invasive and often forms very thick mats on most slow-moving water bodies thereby causing death of flora and fauna in deeper layers of water body. It also causes economic losses to natives who rely on those waters (McConnachie et al., 2003).

LANDSCAPE AND FLORAL INDUSTRY

Ferns are a staple species in the landscape and floral industry. They are valued for their attractive lacy foliage. In the floral industry, ferns are used for both houseplants and greenery for floral arrangements. Some of the most popular ferns for these purposes are sword ferns (*Nephrolepis exaltata*), birdsnest fern (*Asplenium nidus*) one of the rare ferns with solid leaves and floral fern (*Rumohra adiantoides*). These graceful, long-lived perennials have so much to offer! They come in a wide range of colors, shapes, and sizes. Moreover, they are easy to grow and are rarely troubled by diseases or pests. Ferns make an excellent addition to any shady garden and can make as background plantings, or can be interplanted with other shade loving plants such as hostas, astilbes, dicentras and caladiums. They also form an important component of a rockery.

OTHER USES

Food

Ferns are not used as food any more but they did provide some sustenance for the Native Americans. Erect sword fern (*Polystichum munitum*) or licorice fern (*Polypodium glycyrrhiza*) rhizomes are roasted and steamed. Sword ferns taste somewhat bitter and licorice ferns are saccharin-sweet. The unfurling fronds, called fiddleheads, are eaten more often, but anyone attempting to forage for these should note that mature fronds are toxic. Fiddleheads of the ostrich fern (*Matteuccia struthiopteris*), cinnamon fern (*Osmunda cinnamomea*) and vegetable fern (*Diplazium esculentum*) are the varieties eaten.

Fossil Fuels

Ferns were the dominant flora in the landscape many millions of years ago. They existed as big trees e.g., *Cyathea*. When they died, they formed a thick layer of foliage and detritus, along with other trees and foliage. These layers now are deep underground, and have been subjected to bacterial decomposition, heat and pressure for millions of years, eventually turning into coal, which has tremendous economic importance as a non-renewable form of energy.

Fertilizers

Azolla pinnata, commonly known as mosquito fern, is barely recognizable as a fern. It is small and aquatic, floating on ponds and still waters, and has the ability to fix atmospheric nitrogen. *Azolla pinnata* has a symbiotic association with nitrogen-fixing, blue-green algae namely – *Anabaena azollae*. This ability of *Azolla* has led to its extensive use in Asian countries like India, Phillipines and Sri Lanka as biofertilizer with water crops such as rice. By applying *Azolla,* there is increase in total nitrogen, available organic carbon, phosphorous and potassium.

Miscellaneous

Azolla has also been used as a food supplement in fresh or dried or silage form for a variety of animals. Other uses of *Azolla* include hydrogen and biogas production. It is also an ingredient in soap production.

Medicine

Till a decade back the medicinal value of pteridophytes was thought to be less when compared to other archegoniates, the bryophytes and the gymnosperms. Traditional use of ferns for medicinal purposes dates back to Indian systems of medicine. The biomedical system and Ayurvedic systems of medicine, named Sushruta (ca. 100 AD) and Charka (ca. 100 AD), suggested the use of some ferns in the *Samhita* texts. Sanjivni booti, *Selaginella bryopteris* is mentioned in Ramayana with life-saving properties. They are also used by physicians in the Unani system of medicine (Uddin et al., 1998) and have been mentioned in traditional Chinese system of medicine (Kimura & Noro, 1965). In more recent times, ethnobotanical and advanced pharmacological studies have been carried out on ferns and their allies by several investigators (Chen et al., 2005; Chen et al., 2005). The presence of several bioactive substances in ferns and lycopods has generated interest in pteridophytes once again. Among all the pteridophytes examined, taxa from the Pteridaceae, Polypodiaceae, and Adiantaceae exhibit significant medicinal activity.

The leaf and root decoction of commonly occurring *Adiantum lunulatum* syn. *Adiantum philippense* has been found to effective in chest complaints.

The fresh fronds of *Blechnum orientale* are used as a poultice for boils in Malaya; the rhizome is used as anthelmintic in China, eaten during scarcity of food in Malaya, as cure for intestinal worms, bladder complaints in India, and as diaphoretic in Philippines. The rhizome and roots of *Cheilanthes tenuifolia* were used by the tribals as general tonic.

The rhizome of *Dicranopteris linearis*, is used as anthelmintic in Assam while the fronds are used for treating asthma in Madagascar. The rhizome of *Drynaria quercifolia* is used as antibacterial, anti-inflammatory, for treating constipation, diarrhoea, ulcers and other inflammations. The decoction of the plant is used in typhoid fever and fronds are useful in treating swellings. Six different ferns i.e., *Drynaria fortunei, Pseudodrynaria coronans, Davallia divaricata, D. mariesii, D. solida* and *Humata grffithiana* are used as source of medicine commonly known as "Gu-Sui-Bu." The young shoots of *Lygodium flexuosum*, are used as vegetable whereas the rhizome of the plants is boiled with mustard oil and locally applied in rheumatism, sprains, scabies, ulcers, eczema and cuts. The bracken fern - *Pteridium revolutum* syn. *P. aquilinum* has the most varied economic and medicinal uses owing to its wide distribution. The rhizome of the plant is astringent, anthelmintic and is useful in diarrhoea and inflammation of gastric and intestinal mucous membranes. The decoction of rhizome and fronds is given in chronic disorders of viscera and spleen. In times of scarcity the rhizomes by locals are boiled or roasted and eaten or ground into powder for making bread. The rhizomes mixed with malt are used for brewing a kind of beer and are also employed as a feed for stock, especially pigs. The tender fronds of the fern are used as vegetable and also employed in soups. The dried fronds are employed as packaging materials and have also been tried as a source of paper pulp. The bracken fern yields green dye when the fern tops are moderated with alum and copper. The boiled roots of *Pteridium*, which turn black, have been used as chief black pattern material by the Washo, Mono and Yokut Indians.

BIOACTIVE SUBSTANCES IN FERNS AND ALLIES

Spider brake fern (*Pteris multifida*) has various bioactive flavonoids with heat-clearing, antipyretic, detoxification, antibiotic, anti-inflammatory, and antimutagenic activity. The entire plant of *P. multifida* contains various bioactive compounds such as luteolin-7-*O*-glucoside, 16-hydroxy-kaurane-2-β-d-glucoside (Liu & Qin, 2002), luteolin, palmitic acid, and apigenin 4-*O*-α-l-rhamnoside. Certain cytotoxic bioactive compounds such as pterosin sesquiterpenes, have been isolated from *P. multifida* (Shu et al., 2012). Sesquiterpenes form a significant group of secondary metabolites and have a varied range of medicinal activity, including anti-cancer, anti-inflammatory, cytotoxic, plant growth-regulatory, and antimicrobial properties (Baruah et al., 1994).

The octadecanoic acid (a stearic acid) in an ethanol extract of *Pteris tripartite* is hyperprotective. *Pteridium aquilinum*, thought to be a fern with potent anti-cancer properties, contains *p*-coumaric, *p*-hydroxybenzoic, caffeic, ferulic, vanillic, and protocatechuic acids. Some bioactive flavones such as amentoflavone, robustaflavone, biapigenin, hinokiflavone, podocarpusflavone A, and ginkgetin, in *Selaginella* sp. were reported to have antioxidant, antivirus, and anti-cancer activity (Li et al., 2014). The free radical scavenging bioactive principles viz. flavan-4-ol glycosides, abacopterins, huperzine A, and isoquercetin have been derived from a few ferns including *Abacopteris penangiana, Huperzia selago, Equisetum arvense,* and *Dryopteris crassirhizoma.*

The antioxidant activity has been reported for several fern species, namely *Polystichum semifertile, Nothoperanema hendersonii, Braomea insignis* (Chen et al., 2005), some Chinese ferns, *Selaginella* sp., *Polyopodium leucotomus, Equisetum arvense, Equisetum sylvaticum, Abacopteris penangiana, Drynaria fortunei, Pseudodrynaria coronans, Davallia divaricata, Davallia mariesii, Davallia solida, Humata griffithiana, Marsilea quadrifolia, Cheilanthes anceps, Pteris ensiformis Pteris multifida, Pteris tripartita, Dicranopteris linearis* and *Drynaria quercifolia.* The ferns *Pteris nipponica, Polypodium formosanum, Polypodium vulgare, Polypodium fauriei, Polypodium virginianum, Dryopteris crassirizoma, Adiantum monochlamys,* and *Oleandra wallichii* have been shown to have anti-cancer activity. *Angiopteris evecta* and *Selaginella tamariscina* have been shown to have significant antidiabetic activity.

The lycophytes, *Selaginella willdenowii, Selaginella lepidophyla, Selaginella labordei, Selaginella moellendorffii, Selaginella delicatula, Selaginella tamariscina,* and *Sellaginella doederleinii* were shown to have cytotoxic and antimutagenic effects due to the presence of bioflavonoids (Lee et al., 1999). Amentoflavone and ginkgetin, flavonoids found in *Selaginella,* display neurodefensive activity against cytotoxic stress, which suggests their promising usage to cure neurodegenerative diseases such as stroke and Alzheimer's (Kang et al., 2004).

Leaves of *Asplenium bulbiferum* are applied externally to hemorrhoids, and fronds are taken for liver problems. Decoctions from the rhizomes of *Pteridium aquilinum* are drunk as herbal health tea, and the rhizomes of *Woodwardia unigemmata* are used to cure skin diseases. Leaf juices are used as a tea or bath for curing infertility in women, for abdominal pain, constipation and sore throats (Nwosu, 2002). A paste of leaves and rhizomes mixed with lukewarm milk for 30 d is used as an effective tonic. A decoction of roots and fronds of *Drynaria quercifolia* is used as a tonic for the bowels, and aqueous extracts of the rhizomes are applied to wounds and cuts. Dried rhizomes of *P. aquilinum* mixed with milk are used to relieve diabetic disorders, and tender fronds are used as vegetables. Green fronds are used as fodder, make a good soil binder, and are also used in the preparation of manure.

CONSERVATION OF FERNS

The IUCN Red List report shows that despite being known for their resilience, ferns and lycopods continue to be severely affected by human activities, with aquatic species most at risk. The findings reveal that fern and lycopod species are primarily threatened by urbanisation and expanding infrastructure. This leads to the fragmentation and reduction of their habitats. For example, the Dwarf

Moonwort (*Botrychium simplex*) found in several countries including France, Sweden and Austria, is now listed as Endangered as a result of habitat loss through land conversion to forest plantations or tourist developments. Since desert and dryland species are more robust, they are believed to be less threatened. However, destruction of habitat has put survival of fern species at a risk. To top it, the climate change has also posed several strains on the reproductive capacity of the plants.

Another reason for disappearance of certain fern species is pollution. Pollutants from urban and agricultural waste pose a serious threat to many ferns and lycopods. As a result of pollution, many aquatic ecosystems suffer from eutrophication – an increase in nutrients which causes local species to be outcompeted by other native or invasive alien species. One example is *Isoëtes malinverniana* commonly known as Quillwort which has become Critically Endangered. This species is endemic to Italy and has declined by more than 80% in the last 30 years, mainly as a result of pollution through inappropriate irrigation channel management.

Several *in situ* and *ex situ* methods are employed to conserve ferns. The *in situ* methods are based on using the soil from the habitat where ferns are growing as spore bank while for the *ex situ* techniques tissue culture has proved helpful.

BIOLOGICAL INTERVENTIONS

Rhizomes of six different ferns *Drynaria fortunei, Pseudodrynaria coronans, Davallia divaricata, D. mariesii, D. solida* and *Humata grffithiana* are used as source of medicine commonly known as "Gu-Sui-Bu" (Fernández, 2010). This is a cure against inflammation, cancer, ageing, blood stasis, physique ache and bone injuries. Due to the range of treatment, these plants have been over exploited for their fleshy rhizome as a source of "Gu-Sui-Bu", making it mandatory to preserve these species for future generations. Thus, *in vitro* methodology is being investigated to preserve these species. Another fern which has attracted the attention of researchers is the Chinese brake fern (*Pteris vittata*) with an extraordinary ability to tolerate and hyperaccumulate arsenic, up to about 27% of dry weight in its fronds. The brake fern is now being developed for a cost-effective green technology to remediate arsenic-contaminated environments.

The pteridophytes are models for development and differentiation studies. One of the processes involved in sexual development is antheridiogenesis, or the formation of male sex organs, the antheridia. In the fern *Anemia phyllitidis*, antheridiogenesis, a complex process has been prematurely induced by gibberellic acid (GA). The hormone, GA imitates the action of antheridic acid, the main antheridiogen present in *Anemia* sp. It induces many specific cytomorphological and metabolical features in young gametophytes of *Anemia phyllitidis*, during "induction" and "expression" phases of male sex determination. Influence of GA on dispersion of nuclear chromatin, induction of cell division cycle, sugar metabolism and cell wall structural reorganization leads to transverse cell growth and asymmetrical division of antheridial mother cells and to antheridia formatión. Ethylene, plays the role of secondary messenger in these events.

TINY FERNS WITH "SUPER" POWERS

Azolla, is believed to have played a major role in cooling down the planet several million years ago, earning it the title of "super-plant." In what geologists call the "Azolla event," huge blooms of *Azolla* covered the Artic Sea some 50 million years ago. The '*Azolla* event' helped in removing large amounts of carbon dioxide from the atmosphere. This turned a greenhouse Earth with warm temperatures at the poles, into today's planet with polar ice caps.

Azolla is also used as a "green manure" in rice farms in Asia. The fern, sometimes called the mosquito fern or fairy moss, harbors a symbiotic nitrogen-fixing cyanobacterium called *Nostoc azollae* within its leaves. The bacterium in this association captures atmospheric nitrogen converting it into forms that the ferns and rice plants can use. The association of *N. azollae* with *Azolla* is throughout its life cycle and gets transferred from parent to the offspring every time the fern reproduces. However, in most other plants, each generation has to get its symbiotic microbes from the environment. It is also shown that the fern has taken one gene from the bacterium which is a defence against animals grazing on it (Li et al., 2018).

Ferns usually have large genome size which may be as many as 720 pairs of chromosomes and genomes as big as 148 billion base pairs of DNA sequences (Gb) making sequencing difficult. Li and his team, however, found that both *Azolla* and its sister genus, *Salvinia*, have extremely small genomes. While *A. filiculoides* is 0.75 Gb genome, *S. cucullata* is 0.26 Gb, making both good candidates for genome sequencing. The genome sequence for both the ferns and also of cyanobacterium with which they enter into symbiotic relationship for fixing atmospheric nitrogen, have helped to elucidate the mechanism of symbiosis. The secrets of this natural biofertilizer (fern) may help lead to future sustainable agricultural practices (Li et al., 2018).

C-FERN

Ceratopteris richardii, a fern species belongs to the family *Pteridaceae* and is one of several genera of ferns adapted to an aquatic existence. The diploid sporophytes are extremely heteroblastic, initially producing entire sterile leaves and progressing to highly dissected fertile leaves which, under culture conditions, produce many spores continuously throughout the year within sporangia on their enrolled leaf margins. In comparison to many other ferns including *Pteridium*, *Ceratopteris* sporophytes are relatively small, reaching 30–40 cm in height. This feature particularly coupled with its ease of growth in culture has been responsible for the widespread application of *Ceratopteris* as model plant biology to demonstrate plant lifecycles, genetics and development, and in research laboratories.

Ceratopteris richardii, an aquatic fern represents a useful experimental system for studies on plant development. The cellular, molecular and gene expression changes that occur in gametophytes during polarity development, have been investigated in this fern. Responses to the environmental stimuli such as light and gravity have also been studied and molecular approaches are used to characterize proteins, involved in mediating the coupling of light and gravity stimuli to morphogenetic changes in plants. The fern is also considered a good material for genetic transformation. The use of *C. richardii* in genetic research studies has been valuable to understanding fern and plant evolution as a whole, and in 2019 «C-fern» became the first homosporous fern to have its genome partially assembled, thus acting as a reference genome to which other ferns can be compared (Marchant et al., 2019).

Based on this particular species, a patented strain, called "C-Fern", was developed in 1995 as a scientific aid and teaching tool in biology. The use of «C-Fern» is facilitated by the fact that it grows readily in a cell-culture dish on agar media, reaching sexual maturity within 2–3 weeks of *spore* inoculation, with motile sperm cells being visible. Over the course of about 6 weeks germination, sex determination and development of gametophytes, fertilization, embryogenesis, organogenesis, and sporophyte growth can all be observed. This allows comprehensive study of the life cycle of homosporous ferns in a relatively short time period. In addition, due to the small size of the plant many specimens can be observed growing simultaneously, allowing for larger sample sizes in research studies. Following the culture of «C-Fern» in dishes, it can be transplanted to a dirt substrate, where it can be further allowed to grow and future generations can be used for subsequent studies.

Due to the development of C-Fern, research into fern biology has become easy and *C. richardii* has been used as a model organism to study vascular plant cell walls, alternation of generations (and associated mutations), genetics, population dynamics, and the effects of mitotic disrupter herbicides, among other topics (Spiro & Knisely, 2008; Leroux et al., 2013). Despite being genetically identical the inoculated spores can give rise to both hermaphrodites and male gametophytes, depending on the secretion of antheridiogen (Schedlbauer & Klekowski, 1972). This phenomenon has been used to study plant pheromones and the cascade of events that lead to epigenetic changes in ferns. The ability to switch from bisexual to all male spores may provide an evolutionary advantage by promoting outbreeding (Walker, 2012).

The fern cell walls are compositionally similar, though not identical, to those of flowering plants. While the mannose-containing polysaccharides such as mannan and glucomannan appear to be abundant in ferns, pectins appear to be present in lower concentrations than found in other plants (Silva et al., 2011). On the other hand, some wall components structurally and functionally indicate pre-date divergence of ferns from gymnosperms and flowering plants. β-expansins, wall-acting proteins which mediate acid-induced wall creep (McQueen-Mason & Cosgrove, 1995), have been identified from the ferns *Marsilea quadrifolia* and *Regnellidium diphyllum* (both species of aquatic ferns) by their homology to flowering plant. Interestingly, the protein extracts from *M. quadrifolia* are capable of inducing wall creep in cucumber cell walls (Kim et al., 2000). The importance of cell wall composition and metabolism to plants environmental responses and survival, as well as exploitation of them, deem wall composition worthy of extensive exploration.

Finally, ferns existing today represent a genetic inheritance of great value as they include species of ancient vascular plants, with direct connection in the events of the past for settling life on Earth. Ferns have been with us for more than 300 million years, and in that time, the diversification of their form has been phenomenal. The ferns were at the pinnacle during the Carboniferous Period (the age of ferns) and formed the dominant part of the vegetation of that time. Most of the ferns of the

Carboniferous age became extinct, but some survived and later on evolved into the modern ferns. Ferns grow in many different habitats around the world, in all continents except Antarctica and most islands, favouring moist temperate and tropical regions. The life cycle of the ferns has worked quite successfully for millions of years and they can answer many questions related to evolution and development of vascular plants.

COLOUR PLATES

PLATE 1A- J. *LYCOPODIUM*

A&B. Morphology

A – *Lycopodium clavatum*, the plant has a creeping stem with erect tips. Smaller and erect branches at tips bear cones or strobili that contain spores. Basal surface of creeping stems bear dichotomously branched, thin adventitious roots all along their length. B – *Lycopodium phlegmaria* which is an epiphyte on tree trunks and branches, twines tightly with the support. Fertile or the strobilus bearing branches in both species possess scale-like yellowish green leaves while the sterile branches bear linear green leaves.

C-F. Anatomy - Stelar Region (Transverse Section)

C- Stem to show epidermis, a well differentiated cortex (outer, middle and inner) and the stele. D -*Lycopodium* sp – the stelar region enlarged to show actinostele. The xylem is in form of radiating ribs or is star-shaped; it may be modified with the free ends of xylem extended into fan-like structures. In the arms of the star lies phloem. In some species, furrowing of stele may lead to plate-like lobes - plectostele as in *L. volubile* (E) or formation of mesh-like xylem masses, the mixed protostele (F).

G & H. Reproduction

G- Longitudinal Section - *Lycopodium clavatum* to show sporophylls bearing sporangia are arranged around the central axis. The sporophylls possess a small flange on ventral side which extends to protect the sporangium arising on the upper or dorsal (adaxial) surface of the sporophyll (foliar or epiphyllous condition).

H- A magnified view of the same to show kidney-shaped or reniform sporangia containing spores of one kind only (isosporous, homosporous).

I & J. Prothallus

I -The prothallus in *L. cernuum* is small with rhizoids projecting from the undersurface while sex organs are borne on the upper surface. The surface buried under substratum is dark in colour and has endophytic region. J- The prothallus may be top-shaped with a convolute margin (*L. clavatum*), or carrot- shaped (*L. complanatum* and *L. annotinum*). The top of the prothallus is lobed and bears sex organs. While most of the gametophyte is parenchymatous, the tissue differentiation is noted in the lower portion. The central region constitutes storage tissue seen as vertically elongated cells. The radially elongated, closely packed chlorenchymatous cells constitute the palisade mycorrhizal layer.

PLATE 1–*LYCOPODIUM*

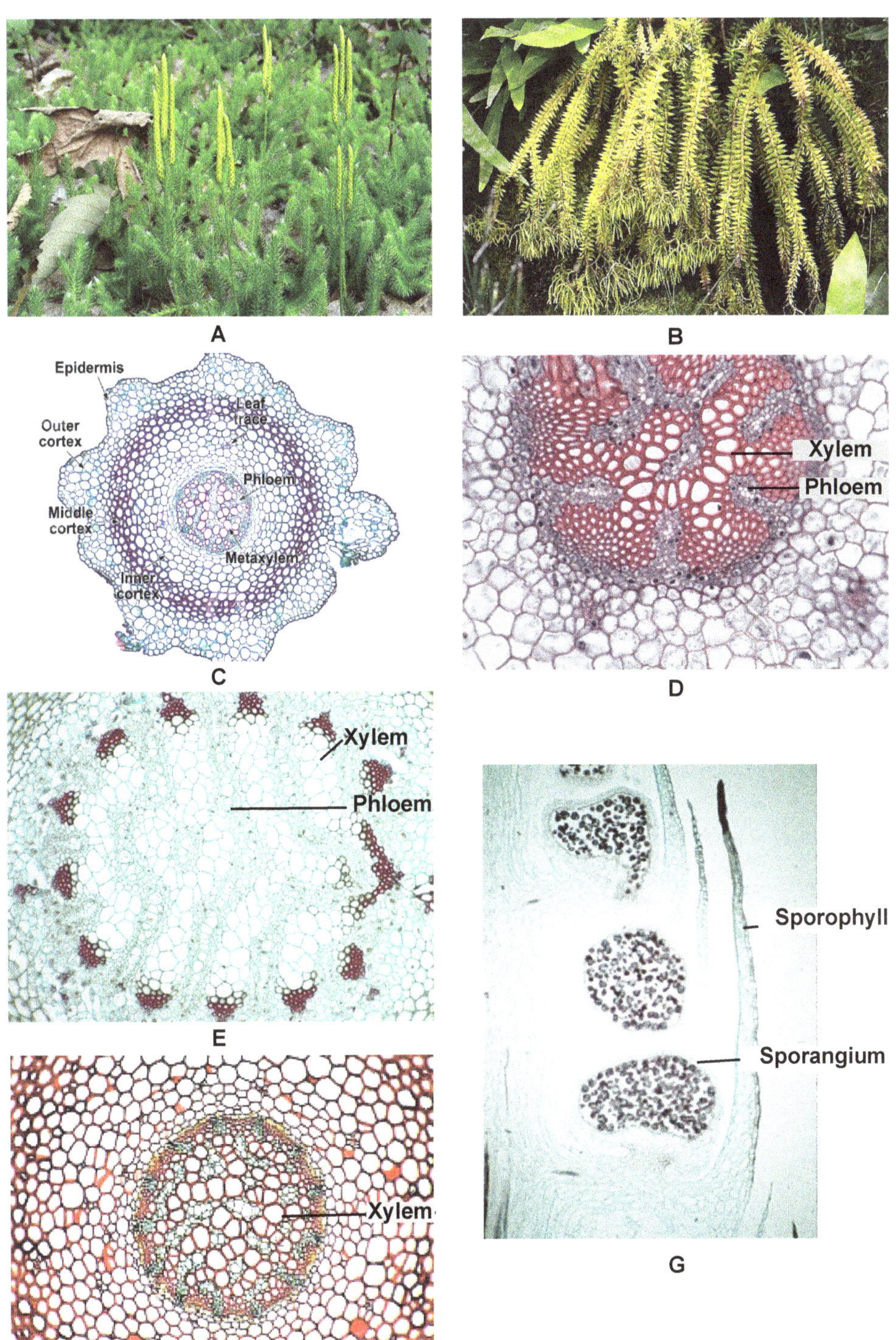

PLATE 1 (CONT...)

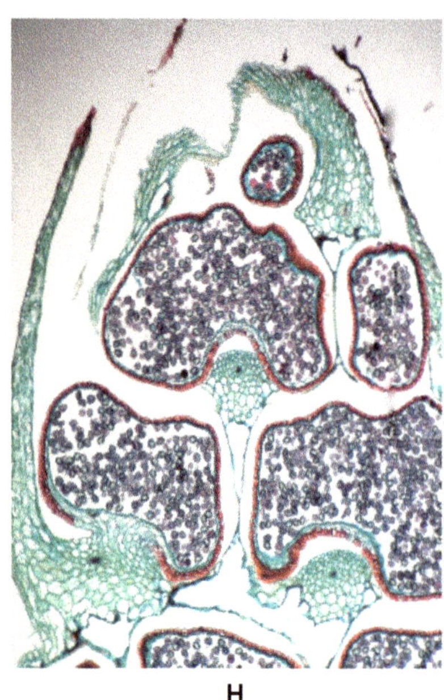

H

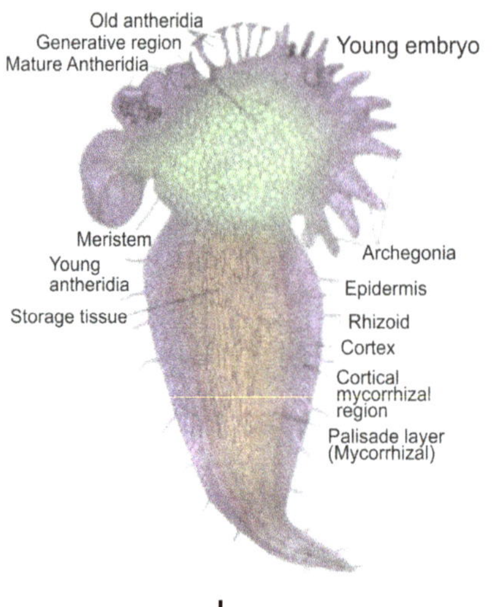

I

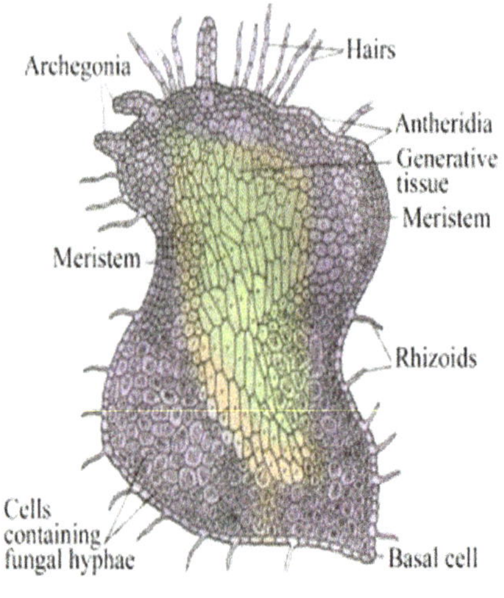

J

PLATE 2A- R. *SELAGINELLA*

A-D. Morphology

A – Plant when prostrate is a sprawling loose mat, with stems elevated on prominent rhizophores; stem is branched with leaves arranged in four ranks.

B- Section of the leaf to show bizonoplast responsible for iridescence. BP (bizonoplast), DE (dorsal epidermis), VE (ventral epidermis), MC (mesophyll cell), IS (intercellular space).

C,D – The leaves (or Sporophylls) are compactly arranged to form strobili (C). The prostrate and pendent species are heterophyllous with two rows of larger leaves and two rows of smaller leaves (D).

E-I. Anatomy (Transverse Sections)

E, F- Stem to show tristelic condition and distinct trabeculae. Stele magnified in (F) to show protostele with metaxylem in the centre forming a solid core and protoxylem groups towards periphery. The xylem core is surrounded by the phloem which remains surrounded by pericycle.

G- Leaf with epidermis, differentiated mesophyll and a central stele.

H- Root with distinct root hairs and a differentiated cortex.

I- Rhizophore which resembles root but lacks root hairs.

J-R. Reproduction by spores

J – Strobili are erect, compact and tapering.

K, L – Whole Mount - Microsporophyll (K) and megasprorophyll (L).

M,N– Whole Mount - Strobili with megasporangia intermixed with microsporangia (M) and with megasporangia placed on the lower portion of the axis while microsporangia are towards apex (N).

O-Q – Longitudinal Section - Strobili to show distribution of megasporangia on one side of the axis while microsporangia occupy the other half of the strobilus (O), the intermingled micro- and megasporangia (P). One megasporangium magnified to show large megaspores with distinct wall (Q).

R – Longitudinal Section - Megagametophyte with compact prothallial cushion that bears archegonia and a large food laden distal part.

PLATE 2A-R. *SELAGINELLA*

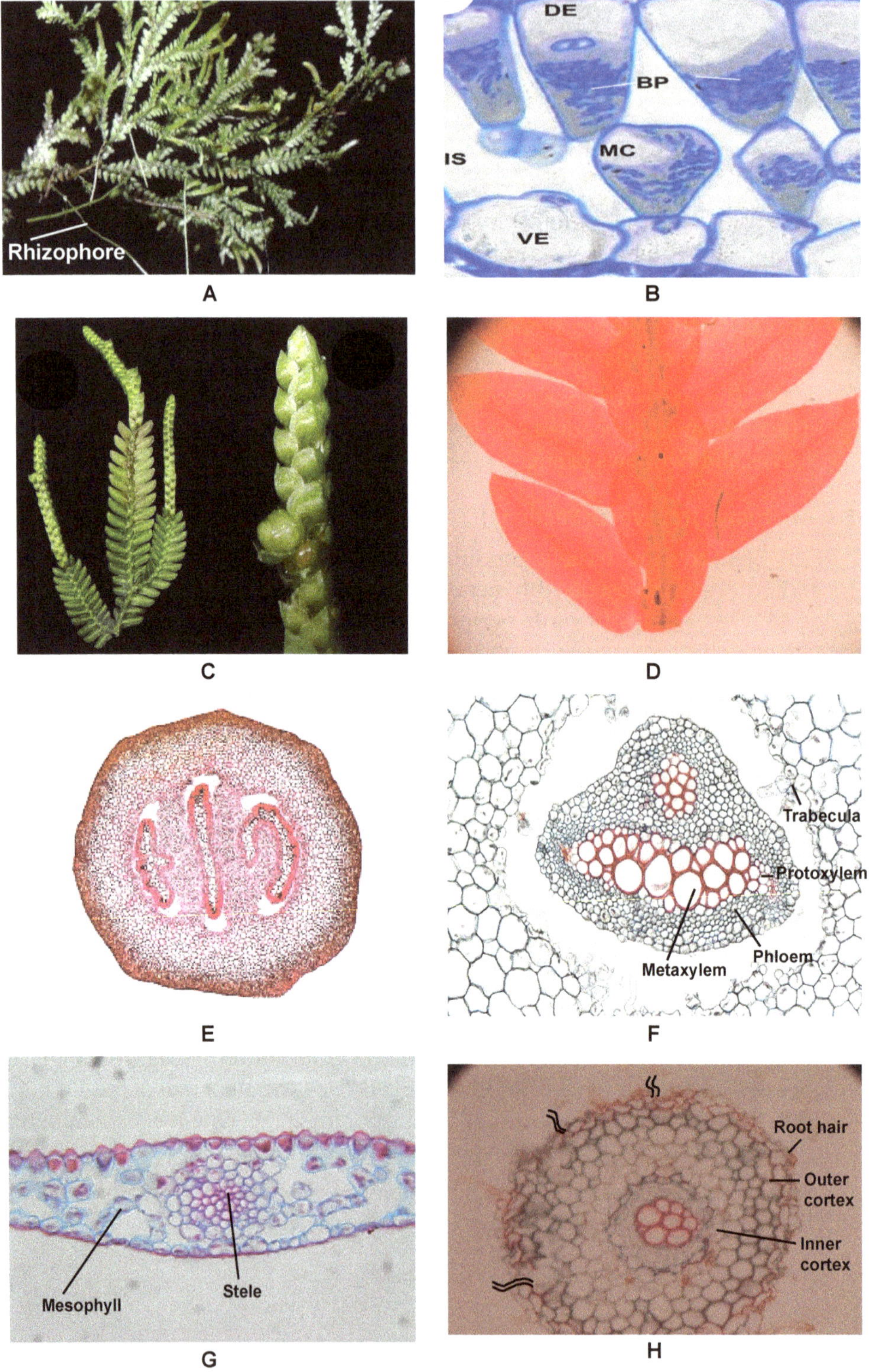

PLATE 2 (CONT..)

Cortex
Stele
I
J
Microsporophyll
Microsporangium
K
Megasporophyll
Megasporangium
L
D
B
E
C
A
M
Microsporophyll
Microsporangium
Megasporophyll
Megasporangium
Megaspore
N
Microsporangium
Megasporangium
O
P
Tapetum
Megaspore
Q
Sex Organs
Storage Zone
R

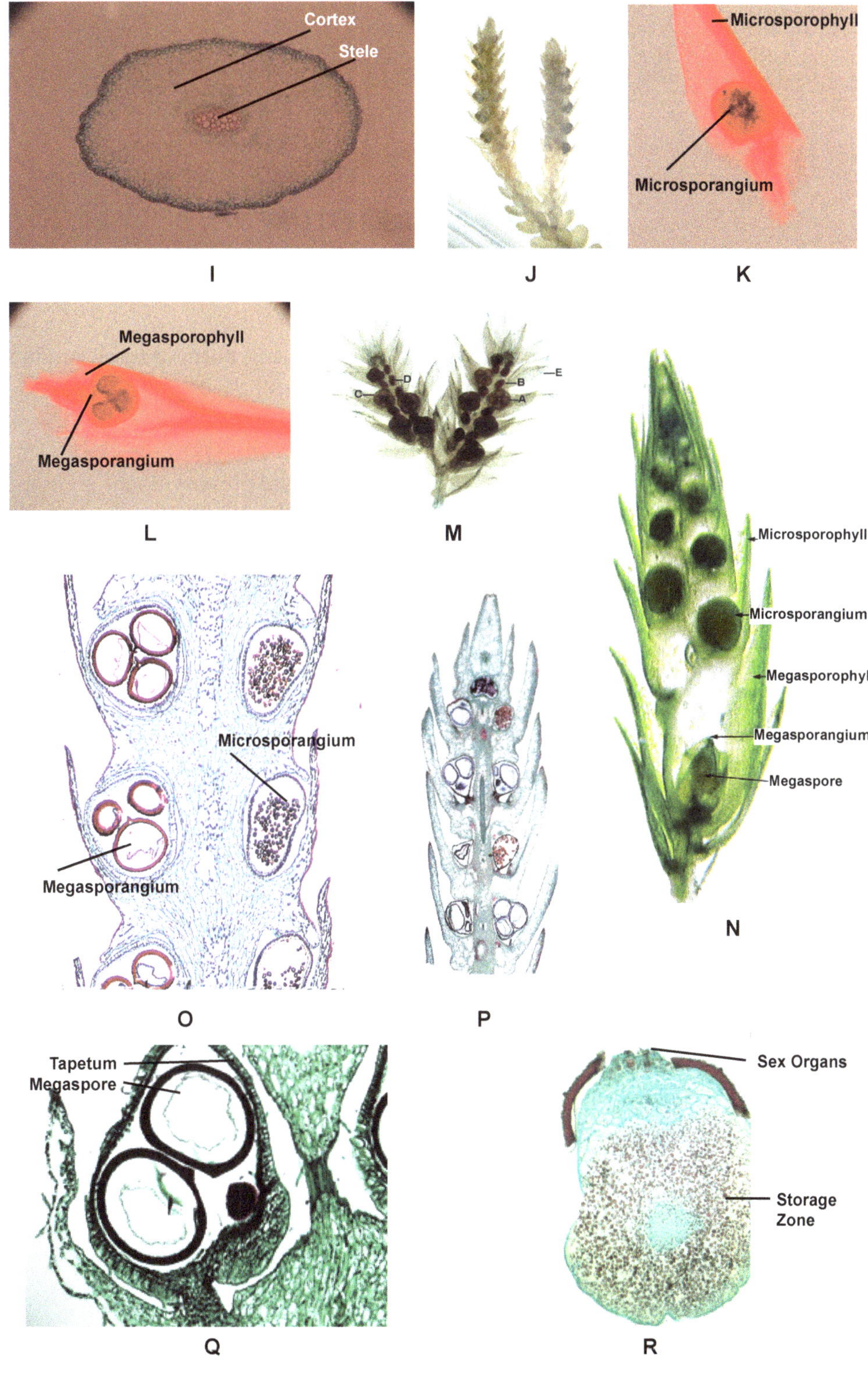

PLATE 3 A-I. *PSILOTUM*

A&B. Morphology

A – *Psilotum* – the whisk fern, in nature. B - Synangia are borne as fused product of three sporangial locules in the axil of bifid appendage.

C&D. Anatomy (Transverse Sections)

C - Aerial stem shows distinct epidermal, cortical and photosynthetic zone and the stele which does not show distinct endodermis and pericycle, is siphonostelic.

D – Rhizome shows darkly stained cells below the epidermis and near the stele which lacks pith and has a centrally placed xylem. The intermediate region of the stem resembles rhizome more but also has chlorenchymatous region.

E-G. Reproduction

E-G - Fertile axis in Longitudinal Section (E) shows the tissues of the aerial stem, noteworthy are the appendages in the axil on which synangia are formed (F). Transverse Section of synangium shows locule with a distinct tapetum around the spores which are all similar. Also seen is the aerial axis that bears the synangia (G).

H & I. Prothallus

The prothallus shows rhizoids and antheridia (H). The dichotomy is distinct and the apices appear whitish (I).

PLATE 3-*PSILOTUM*

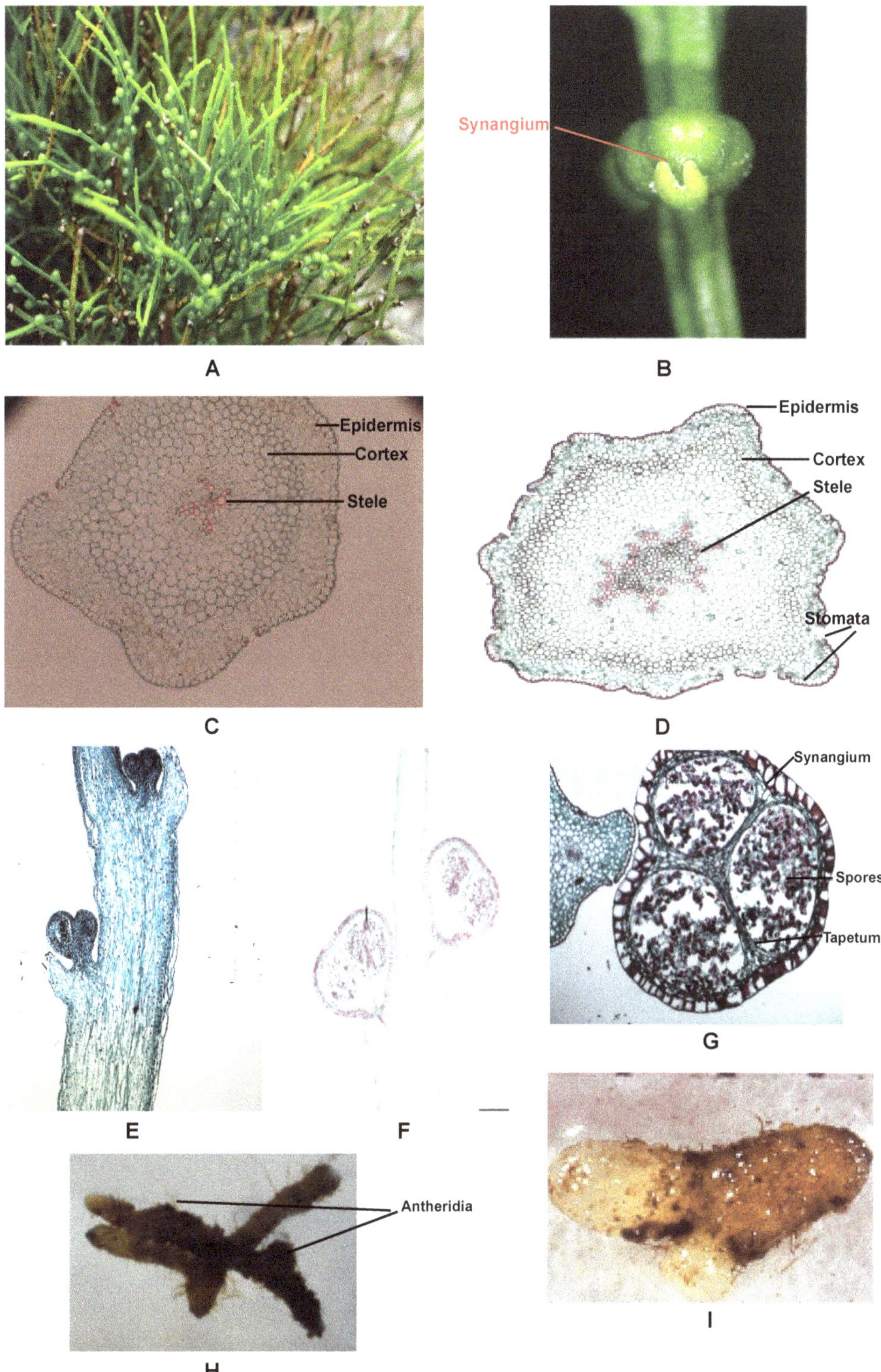

A B C D E F G H I

PLATE 4 A-P. *EQUISETUM*

A. Mature plant with distinct nodes and internodes.

B-F. Anatomy (Transverse Sections)

B- The node shows a parenchymatous pith. C- Internode is hollow with a pith cavity. D- Portion magnified to show ridges with sclerenchymatous patches, vallecular and carinal canals. E-Rhizome with basic plan as the stem but the epidermis is modified with pigmentation and there is not distinction into ridges and furrows. F-Root shows pigmented epidermis with root hairs, cortex and stele with metaxylem in the center and three protoxylem elements towards endodermis which is double layered.

G-N. Reproduction

G, H- An axis bearing a young compact cone in G while a mature cone is loose as seen in H.

I- K- Transverse Sections - Cone.

I- Shows the solid central axis with vascular supply to the sporangiophores.

J- The peltate discs are compactly arranged as the sporangia mature. Once the spores are ready for release the stalk of the sporangiophores lengthens and the discs become loose (K).

L,M- Longitudinal Section cone to show tapering apex in L and the base with bract-like annulus (arrows) in M.

N. Spores with elaters extended under dry conditions.

O & P. Prothallus with erect lobes (O) which bears antheridia at the tips while archegonia occur near the base (P).

PLATE 4- *EQUISETUM*

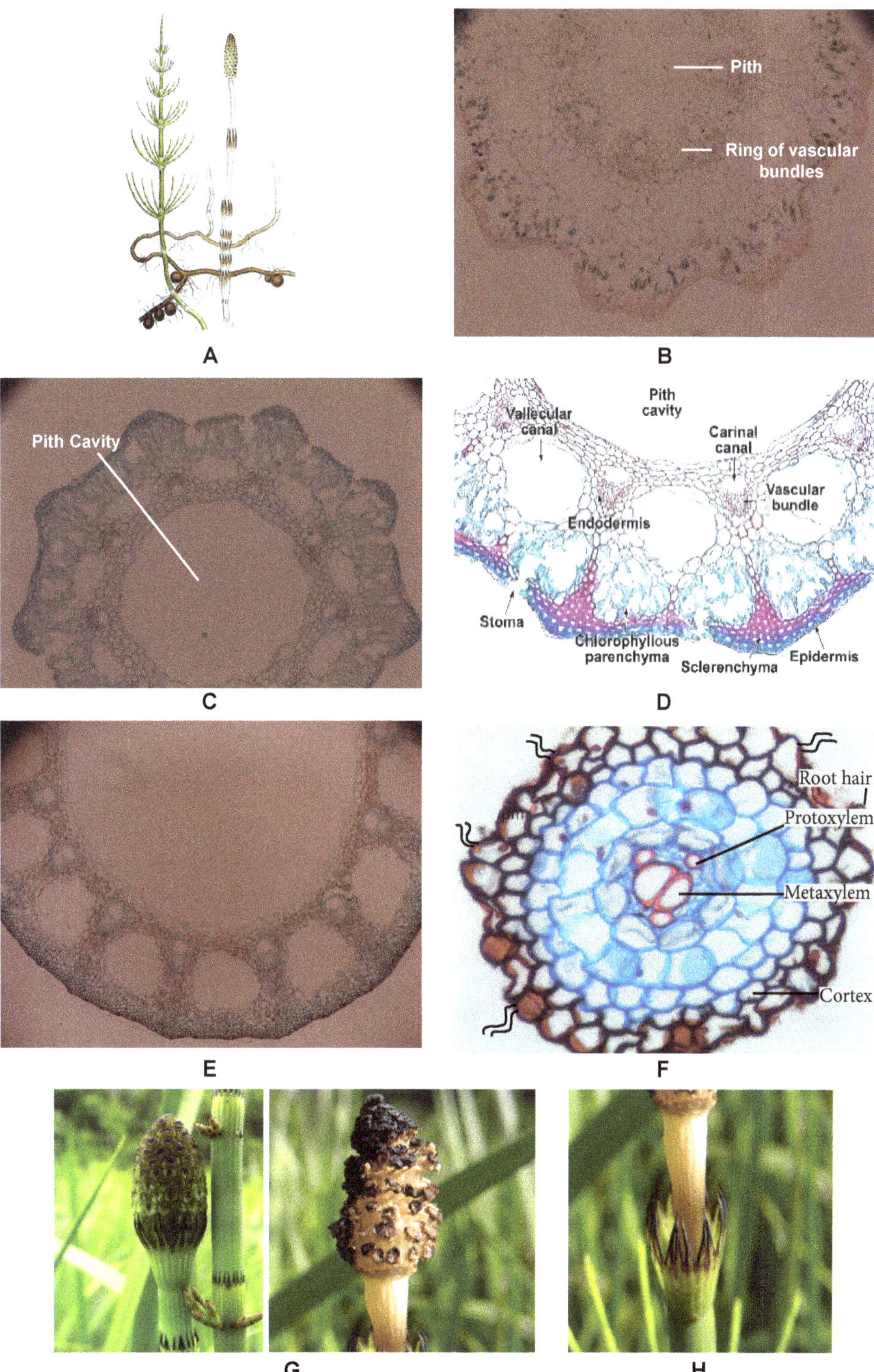

A B C D E F G H

PLATE 4 (CONT...)

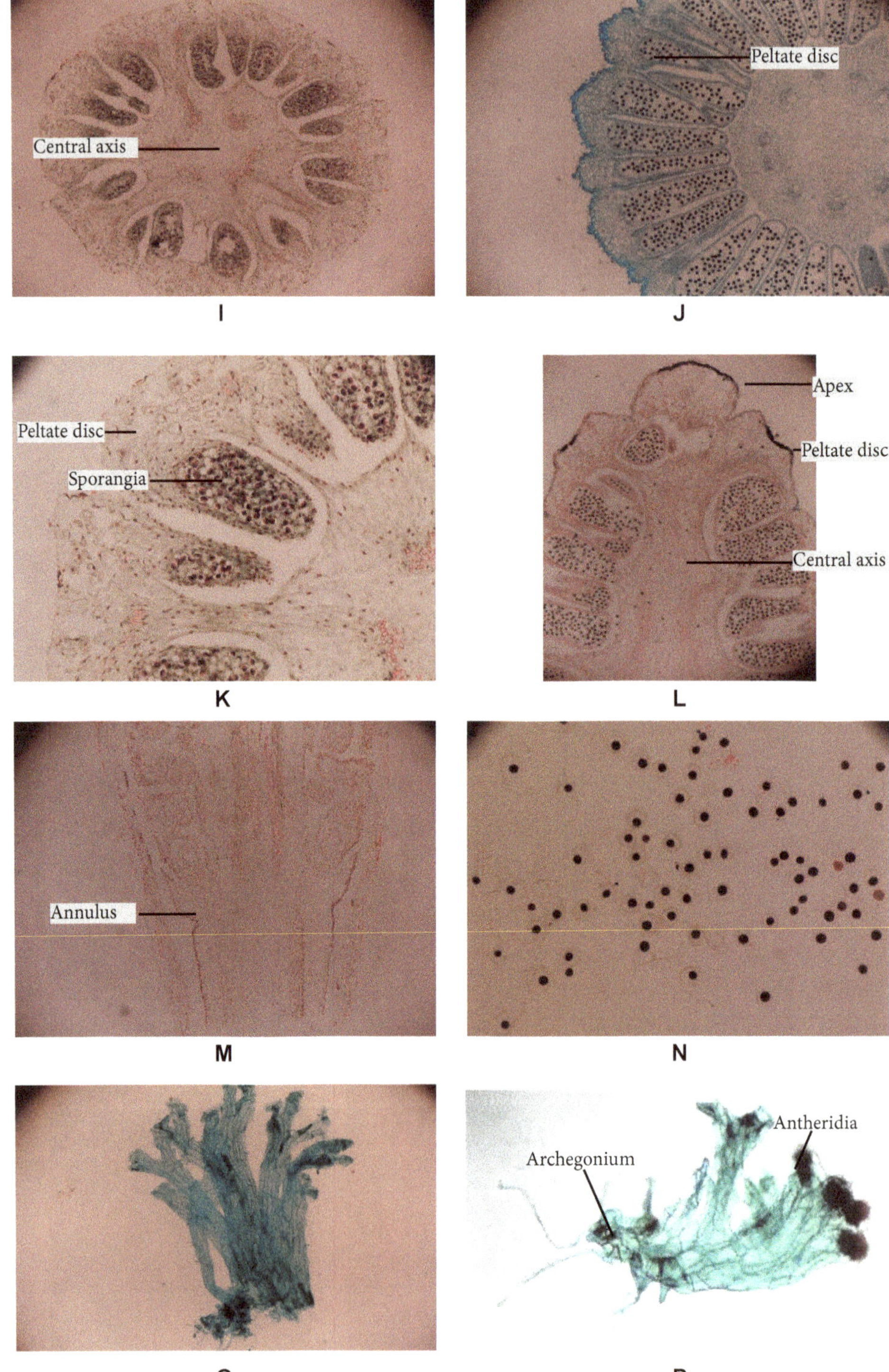

I J K L M N O P

PLATE 5 A-H. *MARSILEA*

A–C. Morphology

A- The plant has a creeping rhizome from which arise leaves with long petioles.

B,C – The quadrifoliate leaves spread out at maturity (B) while they show circinate vernation when young (C).

D. Anatomy

Transverse Section - Rhizome with a well differentiated cortex, aerenchyma and amphiphloic siphonostele.

E-G. Sections of sporocarp

E – L.S to show wall layers and the vascular supply to the sporangia, F,G – H.L.S. To show wall layers, gelatinous pad, indusium and heterospory.

H. Mature female gametophyte with archegonia that protrude into the 'sperm lake'.

PLATE 5- *MARSILEA*

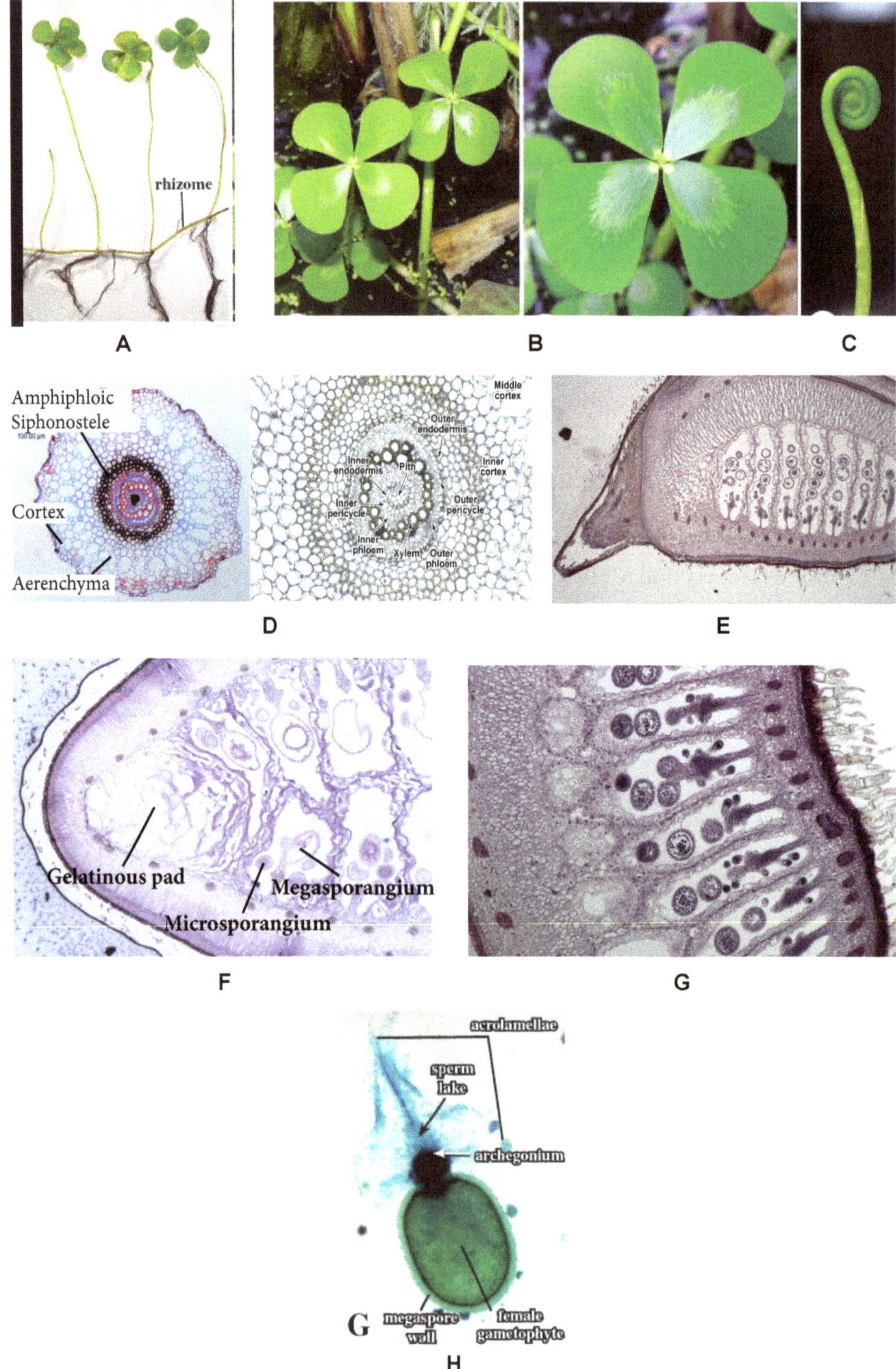

PLATE 6A-T. *PTERIS*

A&B. Habit

The plant grows in clumps and has a beautiful spread.

C-E. Morphology

C – The stout rhizome from which arise fronds, remains covered with roots of secondary origin. D – A sporophyll.

E-The coiled structures, or circinate vernation is an important characteristic of the fern frond.

F-M. Anatomy (Transverse sections)

F- Root showing epidermis with root hairs, broad outer cortex and sclerenchymatous inner cortex which encloses a stele.

G, H- Rhizome showing polycyclic dictyostele (G) with meristeles that contain mesarch xylem (H).

I- Leaf passing through rachis shows stomata in lower epidermis and a well-formed mesophyll.

J- Rachis with V-shaped vascular bundle. The xylem is curved into pronounced hooks that turn inward facing toward the centre of the rachis.

K - Same, the stele may appear jointed with lateral arms taking shape of U.

L- Portion enlarged to show vessels in rachis of pteroid ferns viz. *Ampelopteris prolif*era and *Thelypteris interrupta.*

M- Petiole to show the internal organization of ground tissue and the U-shaped stele.

N- P (Reproduction)

N- The sporophyll bears on the under surface sori which are marginal and continuous.

O-Vertical Section - Sporophyll shows reflexed margin (arrows) forming false indusium.

P - Catapult mechanism of spore release.

Q-T. Prothallus

Q-Prothallus with antheridia, R- prothallus with archegonia. Note the position as the archegonia arise behind the apical cell.

S,T- Prothallus with a young sporophyte.

PLATE 6-*PTERIS*

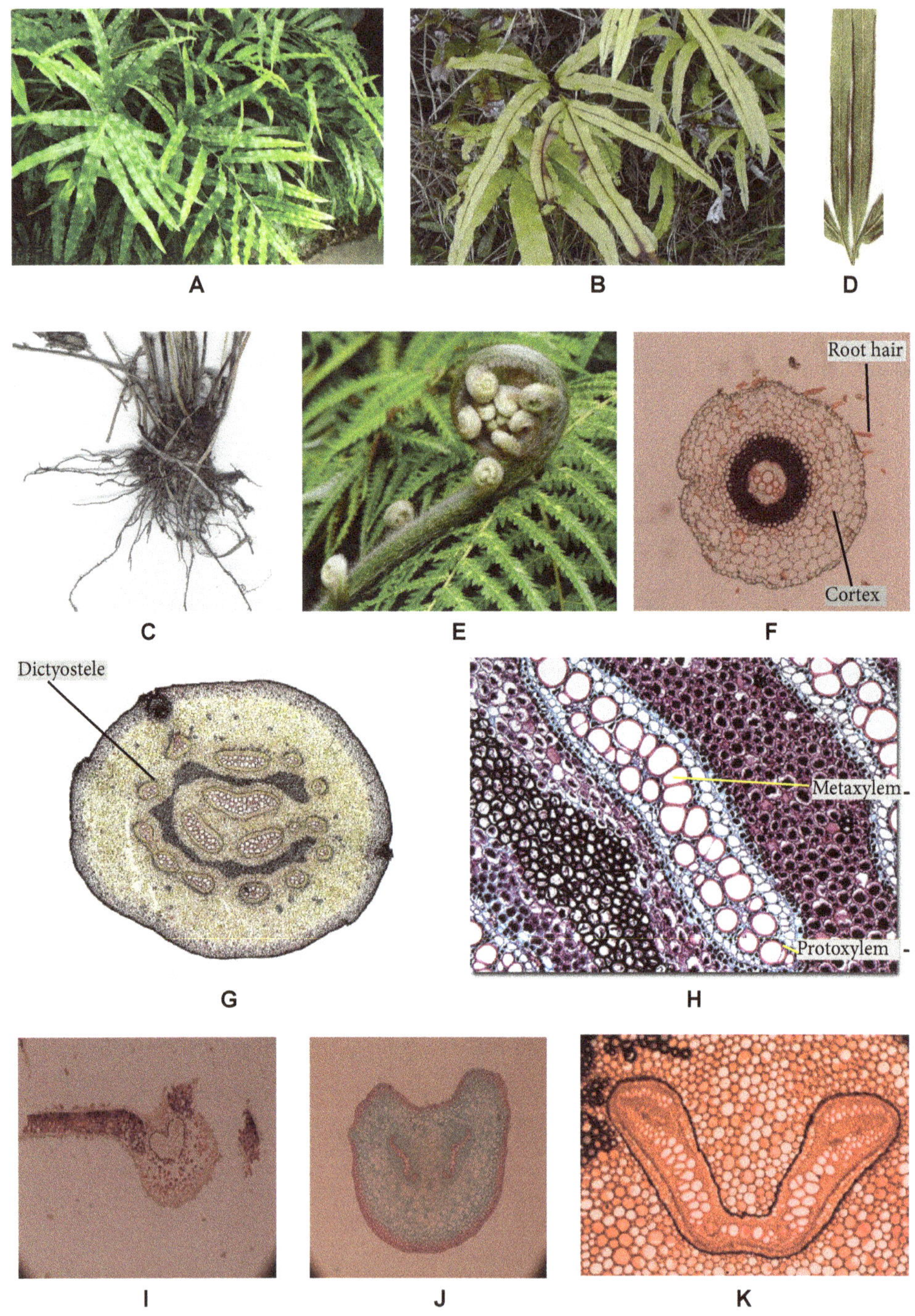

PLATE 6 (CONT...)

L
M
Sori
N
(a)
(b)
(c)
(d)
Annulus
Stomium
O
P
Antheridia
Q
Archegonium
R
S
Spouophyte
Prothallus
T

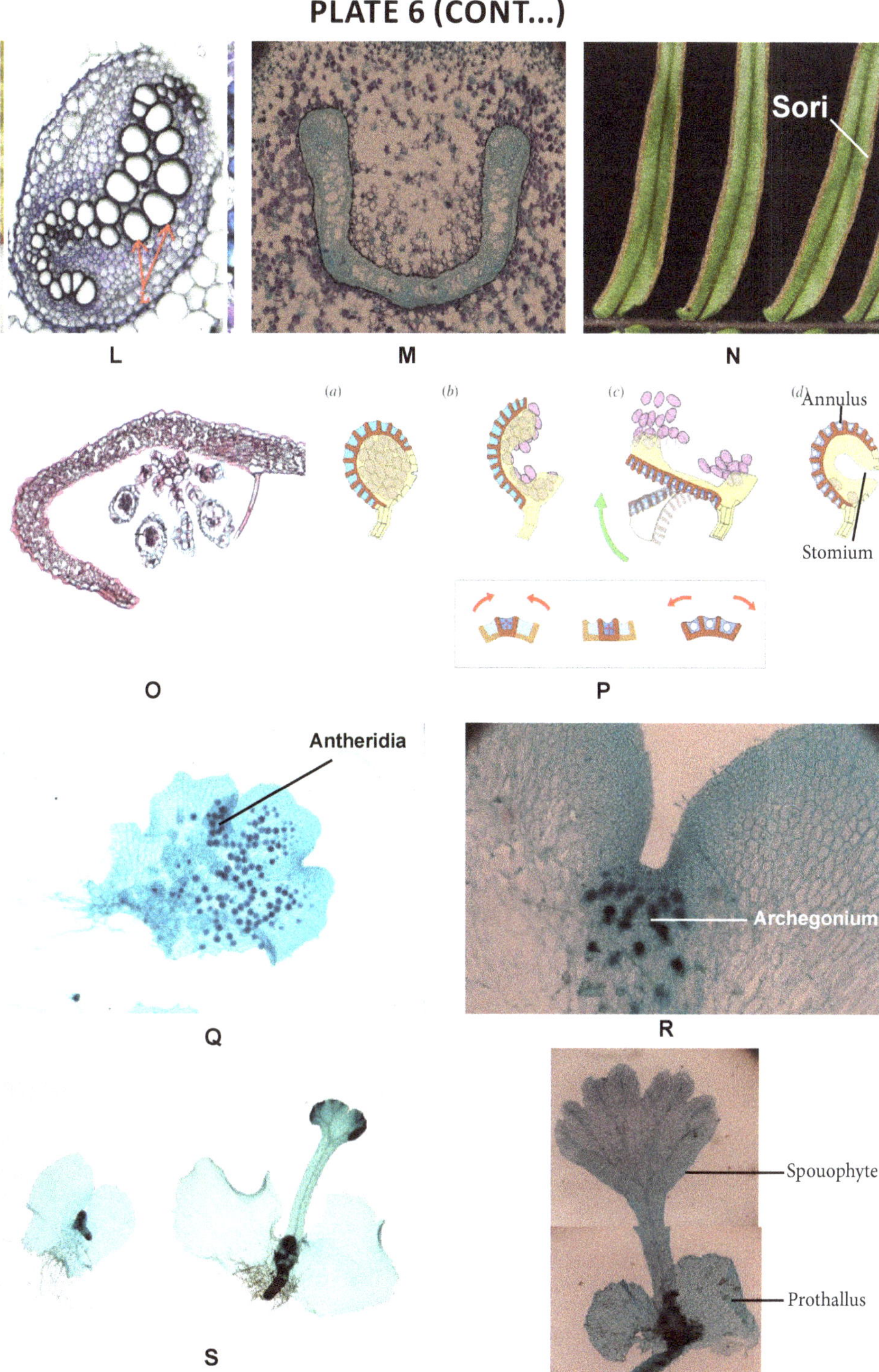

GLOSSARY

Actinostele– Protostele with xylem core having radial arms or ribs, taking up the shape of star. Phloem appears to alternate between the arms of xylem. e.g., *Asteroxylon, Psilotum, Lycopodium serratum.*

Amphiphloic siphonostele– When phloem is present on either side of the xylem. Xylem on inner side has inner phloem, inner pericycle and inner endodermis and xylem on outside has outer phloem , outer pericycle and outer endodermis.e.g., *Marsilea, Adiantum.*

Anisophylly– A condition where the leaves of a pair differ in shape and size.

Atactostele– A type of eustele, found in monocots, in which the vascular tissue in the stem exists as scattered bundles embedded in conjunctive ground tissue. e.g., monocot stem.

Bizonoplast– Unusual chloroplasts in terms of size and thylakoid membrane stacking described in several deep-shade plants. Bizonoplasts are dimorphic in ultrastructure: the upper zone is occupied by numerous layers of 2–4 stacked thylakoid membranes while the lower zone contains both unstacked stromal thylakoids and thylakoid lamellae stacked in normal grana structure oriented in different directions.

Calamitaceae (adj., calamitacean)– A major group of extinct sphenopsids of the Permian Periods, sometimes treated as synonym with Calamitales. It is one of the three families under Equisetales and is characterized by simple linear leaves, secondary growth and arborescent habit, compact terminal fertile regions with whorls of sporangiophores alternating with bracts and that bear spores. They are hypothesized by some workers to have given rise to the Equisetaceae through reduction in plant size, loss of woody tissues and loss of bract whorls within the cones.

Carinal canal– In *Equisetum* and some of its fossil relatives, a number of longitudinal channels formed by the disintegration of the protoxylem remain associated with the metaxylem.

Cauloid– Resembling stem.

Cespitose– Growing in mats, tufts or clumps. Under dry conditions plant rolls into balls.

Dictyostele– A solenostele which is broken due to presence of a large number of leaf gaps into a network of separate vascular strands called dictyostele. Each separated vascular strand is a meristele e.g., *Pteris, Adiantum.*

Ectophloic siphonostele– When phloem is present only on outside of xylem. There is one endodermis and one pericycle external to phloem. Leaf traces may occur but leaf gaps are absent. e.g., *Osmunda.*

Ectophloic solenostele– A solenostele with phloem present on outer side of the xylem.

Enations– Are scaly leaf-like structures or out growths, differing from leaves in their lack of vascular tissue. They are occasionally created by some leaf diseases. On some early plants such as *Rhynia,* they are hypothesized to explain the origin of microphylls in Pteridophytes.

Endoscopic embryogeny– When the embryo develops from the inner of the two cells that results from first division of the zygote. The outer or the external cell develops the suspensor.

Endospory– The production of gametophytes within a spore and is an innovation towards seed habit.

Equisetales (adj., equisetalean)– A major group of sphenopsids that includes extant *Equisetum* and other taxa of the Equisetaceae, as well as the extinct Calamitaceae and, the Archaeocalamitaceae. The order is characterized by whorled, usually single-veined leaves that may be basally fused forming a sheath, and fertile regions consisting of one or more whorls of sporangium-bearing appendages that may or may not alternate with foliar appendages (leaves or bracts).

Euphyllophytes - The clade composed of the monilophytes + seed plants is known as Euphyllophyta, that include almost all living plants, making up more than 99% of living vascular plant species, except the bryophytes and clubmosses. It is by far the largest group of vascular plants, in terms of both individuals and species. The euphyllophytes are characterized by the possession of true leaves («megaphylls»), and comprise one of two major lineages of extant vascular plants. It is a sistergroup of lycophytes.

Eustele– A stele splits into discrete collateral vascular bundles, a modified ectophloic siphonostele. Splitting of the cylinder occurs due to a large number of leaf gaps and the bundles are present in form of a ring with interfascicular areas less distinct. e.g., dicot stem.

Exoscopic embryogeny– Is a type of embryogeny in which principal division of the zygote results in two cells. The external cell frequently shapes the embryo while rest of the plant develops from the inner cell that forms the suspensor or the foot.

Fern allies - It a general term covering a somewhat diverse group of vascular plants that are neither flowering plants nor true ferns, and these plants reproduce by shedding spores to initiate an alternation of generations. Club mosses, horsetails and scouring rushes, and whisk ferns have been variously classified under this category.

Fiddleheads or fiddlehead greens– Are the furled fronds of a young fern, which if left on the plant, each fiddlehead would unroll into a new frond (circinate vernation). Fiddleheads have antioxidant activity, are a source of omega-3, omega-6 fatty acids, are high in iron and fiber, and are therefore used as vegetables. However, certain varieties of fiddleheads have been shown to be carcinogenic. The fiddlehead resembles the curled ornamentation (called a *scroll*) on the end of a stringed instrument, such as a fiddle. It is also called a crozier, after the curved staff used by bishops, which has its origins in the shepherd's crook.

Free sporing– Referring to plants that disperse spores rather retaining them as part of seed production. All land plants other than the seed plants (gymnosperms and angiosperms) are free-sporing.

Haplostele– Protostele with circular xylem, most primitive seen in fossil genera e.g., *Rhynia* and *Horneophyton* and also in some living forms as *Selaginella.*

Heterophyllous species– Having leaves of different types on the same plant as in *Selaginella* where the habit is prostrate e.g., if a species has four rows of leaves, two dorsal rows differ from two ventral rows. The leaves generally differ in size and shape. In homoeophyllous species all the leaves are similar.

Heterospory– The condition in embryophytic plants in which spores are of two types: microspores and megaspores.

Homospory– The condition in embryophytic plants in which all spores produced are of one type, indistinguishable into microspores and megaspores.

Indusia– Is a protective flap of tissue (indusium) attached laterally along the inner side of the sorus in some ferns. It is protective in function and can take a variety of forms. When spores are mature and ready for release, the indusia usually shrivel or bend backwards to expose the sporangia (in the sorus). Occasionally, if the indusia completely cover the sporangia, they may tear irregularly.

Leaf gap (lacuna)– In the stele of many vascular plants, an area of parenchyma cells associated with a leaf trace.

Leptosporangiate ferns – In these ferns sporangia arise from a single epidermal cell and not from a group of cells as in eusporangiate ferns (a polyphyletic lineage). The sporangia are typically covered with a scale called the indusium, which can cover the whole sorus, forming a ring or cup around the sorus, or can also be strongly reduced to completely absent. Many leptosporangiate ferns have an annulus around the sporangium, which ejects the spores.

Meristele– Each of the apparently isolated units of a dictyostele, a distinct vascular strand surrounded by an endodermis that occurs in some ferns and results in the break-up of the main stele.

Mesome– An internode between two forkings that was originally a telome but was relegated to an internodal position by the growth of a new telome.

Mixed protostele– Type of protostele where xylem is divided into several groups and each of the groups is arranged inside a ground mass comprising phloem e.g., *Lycopodium cernuum.*

Monilophytes - Any member of a clade of vascular plants (sometimes called Monilophyta) based on molecular genetic analysis, comprising the whisk ferns, horsetails, ferns, and their allies such as adders' tongues (Ophioglossaceae), moonworts, and grape ferns. Like their sister clade, the seed plants, monilophytes have a differentiated main stem and side branches, but they reproduce by spores instead of seeds.

Phlobaphene– Phlobaphenes are reddish, alcohol-soluble and water-insoluble phenolic substances which can be extracted from plants, or be the result from treatment of tannin extracts with mineral acids.

Phylloid– A leaf or leaf-like structure organogenetically derived by fusion of a system of originally dichotomizing telomes and is used in connection with the telome theory.

Plectostele– A type of protostele where xylem occurs as plates parallel to each other and these alternate with phloem e.g., *Lycopodium clavatum.*

Plesiomorphic state- A primitive or ancestral character or state that was inherited from the ancestor of a group is called plesiomorphy (plesiomorphic character).

Polycyclic stele– A stele with two or more concentric cylinders. It may be polycyclic solenostele or polycyclic dictyostele.

Prothallus– A **prothallus**, or **prothallium**, (from Latin *pro* = forwards and Greek (*thallos*) = twig). The gametophyte of ferns and fern allies. The prothallus develops from a germinating spore. It is a short-lived and inconspicuous (heart-shaped in ferns) structure typically 2–5 millimeters wide, with a number of rhizoids (root-like hairs) growing on lower surface, and the sex organs: archegonium (female) and antheridium (male) on distal surface.

Protocorm– A tuber-shaped body with rhizoids that is produced (also by the young seedlings of various orchids and some other plants) having associated mycorrhizal fungi.

Protostele– Stele with a solid core of xylem, surrounded by phloem, pericycle and endodermis while pith is absent. Simplest and most primitive type is seen in rhizome, stem or roots of many Pteridophytes and has types – haplostele, actinostele, plectostele, mixed protostele.

Protoxylem– Is the part of the primary xylem that develops first, and consists of narrow, thin-walled cells.

Ramenta– Comprises a thin brownish chaffy scales in ferns (also in gymnosperms such as *Cycas)* on leaves and petioles, covers young leaves and is protective.

Resurrection plant– Plants which are able to survive drought, typically folding up when dry and unfolding when moistened.

Rhaphalostachya– A subgenus of *Lycopodium*, where members are prostrate or creeping and there is difference between vegetative leaves and sporophylls.

Rhizome– A continuously growing horizontal underground or subterranean stem which puts out lateral shoots and adventitious roots at intervals.

Rhizophores– One of the downward-growing leafless dichotomous shoots in club mosses of the genus *Selaginella* that bear tufts of adventitious roots at the apex.

Siphonostele– Stele with parenchymatous pith in the center. Pith is interpreted to have two origins: interstellar and extrastelar. Stele is of two types – ectophloic and amphiphloic and according to some workers there is also a sub category - solenostele.

Solenostele– A sub category of siphonostele which is perforated at places due to abundant leaf traces which form leaf gaps. It is of various types – ectophloic solenostele, amphiphloic siphonostele, dictyostele, eustele, atactostele.

Sorophore– When the sporocarp germinates, the hard outer covering splits into halves, allowing the tissue coiled inside to become hydrated. As this internal tissue swells upon imbibing water, it pushes the halves apart through which emerges a long gelatinous worm-like sorophore. The sorophore is a sorus-bearing structure unique to the Marsileaceae which may extend to more than ten times the length of the sporocarp inside which it was coiled. Numerous spore-producing sori remain attached along each side of the sorophore out into the water.

Sperm lake– After imbibing water, the megaspore in *Marselia* releases a gelatinous mass, called *acrolamellae*. The acrolamellae, with apical longitudinal folds and basal horizontal folds, contains a central, liquid-filled region called the *sperm lake*, into which the sperms migrate.

Sphenophyllales– The Sphenopsids are currently classified into two orders, the Sphenophyllales and the Equisetales (Stewart & Rothwell, 1993). The former, a major group of extinct sphenopsids had a single genus *Sphenophyllum*, characterized by whorls of cuneate to broadly ovate leaves with dichotomizing veins, exarch protosteles, secondary growth and fertile regions consisting of whorls of sporangium-bearing appendages with foliar parts. Sphenophyllales were small shrubs or vines that had worldwide distribution, primarily from the Mississippian through the Triassic Periods.

Sphenopsida (adj., sphenopsid)– Major clade of euphyllophytes characterized by jointed stems bearing whorled appendages. The Sphenopsida extend from the Late Devonian through the recent and include the Archaeocalamitales, Sphenophyllales and Equisetales, along with several other extinct lineages.

Sporangiophore– The sporangium-bearing appendage of the sphenopsids. The homology of sporangiophores in different sphenopsid lineages is disputed.

Sporocarps– The sporangia in *Marselia* are borne in special type of organ called sporocarp which are borne laterally on the short and lateral branches of the petiole of leaf either near the base or a little higher up. They are interpreted as modified pinnate leaves or pinnae, and are reniform bearing two halves, each of these with several rows of internal sori.

Stomium– The thin-walled cells of the annulus marking the line or region of dehiscence of a fern sporangium that ruptures to release the spores. It is also seen in flowering plants where it represents the opening in an anther usually between locular cells through which dehiscence occurs.

Strobilus (pl. strobili)– Aggregation of sporangium-bearing appendages (e.g. sporophylls, sporangiophores) at the tip of a shoot with determinate growth.

Synangium– Synangia are characteristic of the ancient group Psilotophyta. They are trilocular or three chambered, spore bearing structure in *Psilotum*. Each synangium is considered as a fusion product of three sporangia.

Telome– Is a hypothetical plant structure in telome theory of the evolution of leaves and sporophylls in vascular plants that consists of one of the vegetative or reproductive terminal branchlets of a dichotomously branched axis.

Trabeculae– A trabecular, plural trabeculae, derived from Latin for 'small beam' is a small, often microscopic, tissue element in the form of a small partition, that supports or anchors a framework of parts (stele) within a body or organ e.g., stem.

Tracheids– They are one of the two types of tracheary elements of vascular plants (The other being the vessel elements); they lack perforations.

Urostachya– A subgenus of *Lycopodium,* where members possess branched or pendant but never creeping stems. The vegetative and sporophylls are similar in appearance.

Vallecular canals (cortical canal)– In *Equisetum* and some of its fossil relatives, a number of large, air-filled, intercellular channels run the length of each internode and are situated approximately between the vascular bundles.

SELECTED READINGS

Chase, M.W. and Reveal, J.L. 2009. A phylogenetic classification of the land plants to accompany APG III. *Botanical Journal of the Linnean Society,* **161**:122-127.

Christenhusz, M. and Chase, M. 2014. Trends and concepts in fern classification. *Annals of Botany.* **113**: 571-594.

Foster, A.S. and Gifford, E.M. 1962. Comparative Morphology of Vascular Plants. Vakils, Feffer and Simons.

Fernandez, H. 2010. Introduction to Working with Ferns In: Fernandez H., Kumar A., Revilla M.A. (ed). Working with Ferns: Issues and Applications. Springer.

Harrison, J. and Morris, J. 2017. The origin and early evolution of vascular plant shoots and leaves.Philosophical Transactions of the Royal Society B. *Biological Sciences.* 73

Kim, J.H., Cho, H.T., and Kende, H. 2000. β-Expansins in the semiaquatic ferns *Marsilea quadrifolia* and *Regnellidium diphyllum*: evolutionary aspects and physiological role in rachis elongation. *Planta.* 212: 85–92.

Li, F.W., Brouwer, P., Carretero-Paulet, L. *et al.* 2018. Fern genomes elucidate land plant evolution and cyanobacterial symbioses. *Nature Plants.* **4:** 460–472.

Leroux, O., Eeckhout, S., Viane, R.L. and Popper, Z.A. 2013. *Ceratopteris richardii* (C-fern): a model for investigating adaptive modification of vascular plant cell walls. *Front Plant Sci.* 23;4:367.

Marchant, D.B., Sessa, E.B., Wolf, P.G. *et al.* 2019. The C-Fern (*Ceratopteris richardii*) genome: insights into plant genome evolution with the first partial homosporous fern genome assembly. *Sci Rep* **9:** 18181. *https://doi.org/10.1038/s41598-019-53968-8*

McQueen-Mason, S., and Cosgrove, D. J. 1995. Expansin mode of action on cell walls: analysis of wall hydrolysis, stress relaxation, and binding. *Plant Physiol.* 107: 87–100.

Spiro, M.D. and Knisely, K.I. 2008. Alternation of generations and experimental design: a guided-inquiry lab exploring the nature of the her1 developmental mutant of *Ceratopteris richardii* (C-Fern). *CBE life sciences education,* 7(1), 82–88.

Schedlbauer, M. D., and Klekowski, E. J. Jr. 1972. Antheridogen activity in the fern *Ceratopteris thalictroides* (L.) Brongn. *Bot. J. Linn. Soc.* 65: 399–413.

Silva, G.B., Ionashiro, M., Carrara, T.B., Crivellari, A.C., Tiné, M.A., Prado, J., Carpita, N.C. and Buckeridge, M.S. 2011. Cell wall polysaccharides from fern leaves: evidence for a mannan-rich Type III cell wall in *Adiantum raddianum*. *Phytochemistry*. 72(18):2352-60.

Schuettpelz, E. *et. al.*, 2016. A community-derived classication for extant lycophytesand ferns. The Pteridophyte Phylogeny Group. *Journal of Systematics and Evolution*. **54**: 563-603.

Walker, T. 2012. *Plants: A Very Short Introduction*. Oxford University Press, p. 49.

VIDEOS

https://www.youtube.com/watch?v=c4YtOT0Z6Ek

https://www.youtube.com/watch?v=c4YtOT0Z6Ek

https://www.youtube.com/watch?v=JKgXX8MwQrY

https://www.youtube.com/watch?v=bnfkDI1xnrk

https://www.youtube.com/watch?v=QecSHBOX1MU

https://www.youtube.com/watch?v=LexTbufxWKU

INDEX

I

J

L

M

www.ingramcontent.com/pod-product-compliance
Ingram Content Group UK Ltd.
Pitfield, Milton Keynes, MK11 3LW, UK
UKHW021010290726
14059UKWH00001BA/61

9 789359 191881